Switching Systems and Applications

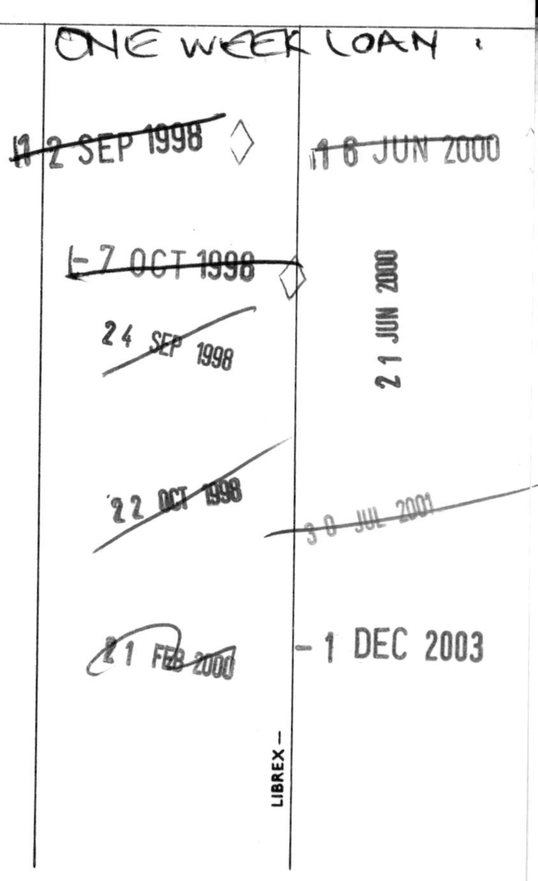

Switching Systems and Applications

Edited by

Fraidoon Mazda
MPhil DFH CEng FIEE

With specialist contributions

Focal Press
An imprint of Butterworth-Heinemann
Linacre House, Jordan Hill, Oxford OX2 8DP
A division of Reed Educational and Professional Publishing Ltd

A member of the Reed Elsevier plc group

OXFORD BOSTON JOHANNESBURG
NEW DELHI SINGAPORE MELBOURNE

First published 1996

British Library Cataloguing in Publication Data
Mazda, Fraidoon F
Switching Systems and Applications
I. Title
621.382

ISBN 02405 1456 4

Library of Congress Cataloguing in Publication
Mazda, Fraidoon F.
Switching Systems and Applications/Fraidoon Mazda
p. cm.
Includes bibliographical references and index.
ISBN 02405 1456 4
1. Telecommunications. I. Title
TK5101.M37 1993
621.382–dc20

92-27846
CIP

Printed and bound in Great Britain by
Biddles Ltd, Guildford and King's Lynn

Contents

Preface

Switching forms the vital hub of any telecommunications system, and the invention of the first electronic switch by Strowger is a landmark in the history of telecommunications technology. This book looks at the principles of switching and describes the construction and applications of public and private switching systems.

The principles of switching circuits are described in Chapter one, from the early analogue electronic exchanges to the modern day digital switches. Chapter two then looks at the various signalling systems which are so closely associated with these switches.

The construction of some of the more popular modern public switching equipment is described in Chapter three. This is followed, in Chapter four, by a description of Private Automatic Branch Exchanges (PABX) and key systems, as used in private networks.

Some of the applications of switching systems are then covered in the following chapters. Chapter five describes the use of Centrex; Chapter six covers call management; the principles of voice processing are explained in Chapter seven; and Chapter eight describes electronic data interchange (EDI). The final chapter of the book deals with the terminal equipment commonly used with switching systems, i.e. telephones and headsets.

Fourteen authors have contributed to this book, all specialists in their field, and the success of the book is largely due to their efforts. The book is also based on selected chapters which were first published in the much larger volume of the *Telecommunications Engineers' Reference Book*, now in its sixth edition.

Fraidoon Mazda
Bishop's Stortford
April 1996

List of contributors

Steve Berrisford
Northern Telecom
(Section 4.1)

Paul W Bizzell
AT&T EasyLink Services
(Chapter 8)

D G Bryant
DFH CEng FIEE
Telecommunication Consultant
(Chapter 1 and Sections 3.2–3.4)

David M Davidson
Northern Telecom
(Sections 4.2–4.7 and Sections 9.1–9.10)

John Holdsworth
Newbridge Networks Ltd.
(Chapter 7)

Fred Howett
Northern Telecom
(Section 4.8)

Kanagendra
B Tec MBA CDipAF
Northern Telecom Europe
(Chapter 5)

Fraidoon Mazda
MPhil DFH CEng MIEE
BNR Europe Ltd.
(Chapter 10)

Malcolm A Nugent
GN Netcom (UK) Ltd.
(Sections 9.11–9.16)

Jean R Oliphant
Rockwell International
(Chapter 6)

Kjell Persson
Ericsson Ltd
(Sections 3.8–3.10)

M Smouts
Alcatel Bell Telephone
(Sections 3.5–3.7)

Eur Ing S F Smith
BSc (Eng) CEng FIEE
Northern Telecom Europe
(Chapter 1 and Sections 3.2–3.4)

Samuel Welch
OBE MSc (Eng) CEng FIEE
Telecommunications Consultant
Formerly Head of Signalling, BT
(Chapter 2)

1. Principles of switching circuits

1.1 Introduction

The simplest form of telephone system consists of two terminations and a metallic connection between them. However, this arrangement is clearly not viable for more than a very small number of terminals and the concept of switching is introduced, whereby each terminal is connected to a central location (exchange) where means is provided to interconnect any two of the total population of terminations.

Within an exchange, more than one connection will be required at any one time but not all terminations will be wanting service at the same time. Practical systems are designed to provide a number of links which is less than the number of terminations determined on a statistical basis. Figure 1.1 shows such an arrangement, based on a matrix of crosspoints where each crosspoint is some form of on/off switch such as a manually operated key, an electromechanical relay or a transistor.

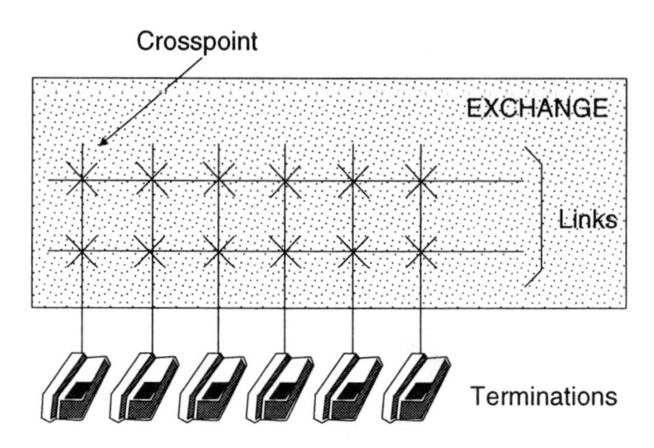

Figure 1.1 Switching principle

This simple example can be regarded as a single node network with all connections passing through one single exchange. This is clearly unreasonable on a global scale and in a practical network, as the number of terminations increases, so must the number of nodes or exchanges, normally on a geographical basis. There is then a need to provide interconnections (junctions) between the different exchanges. Each exchange must be able to handle connections between its own terminations and between them and the external junctions. In large networks it is usual to transit switch calls between other exchanges and some exchanges exist solely for this purpose and have no terminations of their own. This leads to a hierarchical arrangement of exchanges as shown in Figure 1.2.

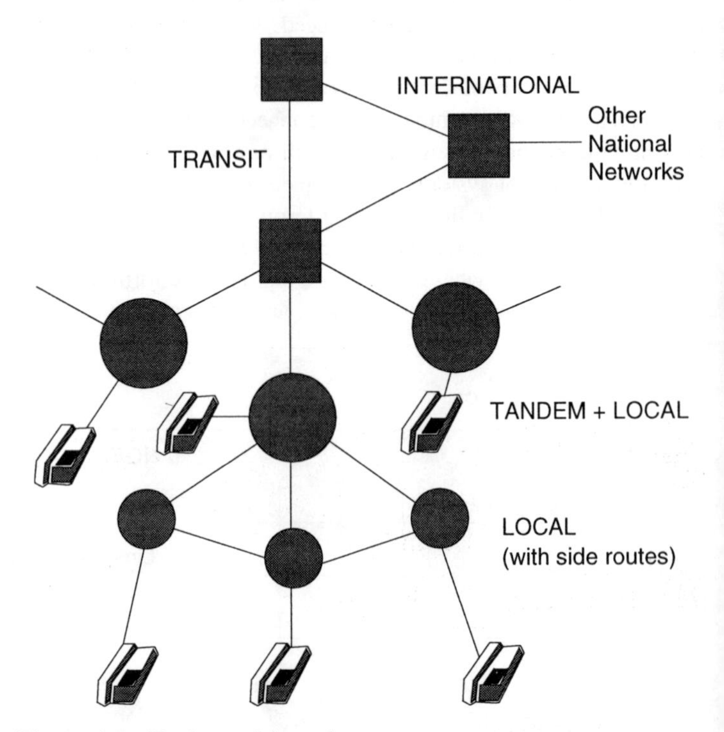

Figure 1.2 Exchange hierarchy

The various types of call routing are shown in Figure 1.3 and traffic is not usually evenly divided over these types of call. Only a small proportion (5% to 15%) of the traffic is normally between customer terminations (subscribers) on the same exchange. The remainder is divided between incoming and outgoing calls. Services such as emergency, speaking clock etc. are usually centralised at one exchange in an area, so that the outgoing traffic at other exchanges normally exceeds the incoming traffic. The proportion of transit (tandem) traffic varies from zero at small rural exchanges to 100% at exchanges provided solely for this purpose.

A network must, therefore, consist of terminations provided at the customers' premises, a transmission medium to connect them to the local exchange and a means of signalling between the termination and the exchange. As well as the telephone customers, there is an increasing requirement for the connection of teleprinters and other data terminals and the public switched telephone network (PSTN), which was originally established to carry voice traffic, is increasingly being used to carry data traffic and to provide a variety of services beyond plain old telephone service (POTS).

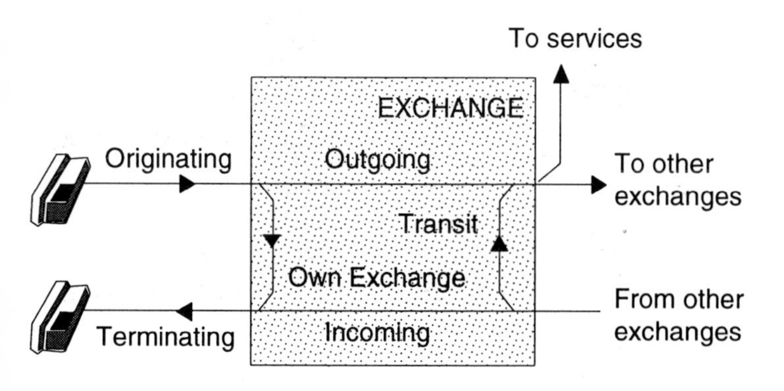

Figure 1.3 Call routing

1.2 Structure of telephone systems

1.2.1 Telephone instrument

The speech transmission elements of a telephone instrument, (see Figure 9.1) consist of a receiver and transmitter (microphone) usually assembled in a common mounting to form a handset and connected to the line through an anti-sidetone induction coil (hybrid transformer) or an equivalent electronic circuit. The complete instrument also incorporates signalling elements, the exact form of which depends upon the type of exchange with which it is designed to interwork. Generally they comprise a dial or push buttons and associated gravity switch for signalling to the exchange and a ringer (bell) or an electronic detection circuit and sound transducer for receiving calling (ringing) signals from the exchange.

1.2.1.1 *Transmitter*

Most instruments employ a carbon microphone (see Figure 9.2). Sound waves cause the diaphragm to vibrate, varying the pressure exerted on the carbon granules. This produces corresponding variations in the electrical resistance between the granules and hence in the current flowing through them when connected to a d.c. supply.

Power may be derived from a battery or other d.c. supply locally at the instrument but in public systems it is now the universal practice for a single central battery to provide a supply to all lines on the exchange. The a.c. component of the line current produced in this way is known as the speech current and the d.c. component derived from the battery is the microphone feed current (transmitter feed) or polarising current.

The feed to each line is decoupled by individual inductors, often in the form of relay coils, and speech is coupled from one line to another through the exchange switching apparatus by means of capacitors or through a transformer. The combination of d.c. feed, battery decoupling and speech coupling components is known as a transmission bridge or feeding bridge. The relays providing the battery decoupling also serve to detect line signals to control the holding and releasing of the connection.

The microphone resistance is nonlinear and is subject to wide variations due to temperature, movement of granules and, of course, the effects of sound. Typically it is between 40 and 400 ohms under normal working conditions.

Some telephone instruments employ other types of microphone (e.g. moving coil) which do not themselves require a d.c. feed. Such microphones, however, lack the inherent amplification of the carbon microphone. They are usually supported by a transistor amplifier driven by the line current and designed to operate with similar feeding bridge conditions to carbon microphone instruments.

1.2.1.2 *Receiver*

Most types of receiver operate on the moving-iron principle employed by Alexander Graham Bell for both transmitting and receiving in his original telephone in 1876. In some cases the diaphragm is operated on directly by the magnets but in more efficient arrangements the design of armature and diaphragm can each be optimised for its own purpose without conflict of mechanical-acoustic and magnetic requirements. A permanent magnet is included in the simple receiver to prevent frequency doubling due to the diaphragm being pulled on both the positive and the negative half cycle. In the rocking armature arrangement (see Figure 9.3) no movement of the armature would occur at all without the permanent magnet.

Unlike the carbon microphone, the moving iron receiver requires no direct current for its operation and is usually protected against possible depolarisation by a series capacitor.

1.2.1.3 *Anti-sidetone induction coil (ASTIC)*

This is a hybrid transformer which performs the dual function of matching the receiver and transmitter to the line and controlling sidetone.

Sidetone is the reproduction at the receiver of sound picked up by the transmitter of the same instrument. It occurs because the transmitter and receiver are both coupled to the same two-wire line. The most comfortable conditions for users are found to be when they hear their own voice in the receiver at about the same loudness as they would hear it through the air in normal conversation. Too much sidetone causes talkers to lower their voice, reduces their subsequent listening ability and increases the interfering effect of local room noise at the receiving end. The complete absence of sidetone, however, makes the telephone seem to be 'dead'.

The use of a hybrid transformer divides the transmitter output between the line and a corresponding balance impedance. If this impedance exactly balanced that of the line there would be no resultant power transferred to the receiver. In practice there is some transfer to produce an acceptable level of sidetone.

1.2.1.4 *Gravity switch (switch hook)*

To operate and release relays at the exchange and thus indicate calling and clearing conditions, a contact is provided to interrupt the line current. This contact is open when the handset is on the rest and closes (makes) when the handset is lifted. Because it is operated by the weight of the handset it is known as the gravity switch. It is also known as a switch hook contact and the terms off-hook and on-hook are often used to describe signalling conditions corresponding to the contact closed or open respectively.

1.2.1.5 *Dial*

This is a spring-operated mechanical device with a centrifugal governor to control the speed of return. Pulses are generated on the return motion only, the number of pulses up to ten depending upon how far round the finger plate is pulled.

The pulses consist of interruptions (breaks) in the line current produced by cam-operated pulsing contacts. A set of off-normal contacts operate when the finger plate is moved and are used to disable the speech elements of the telephone during dialling. These dial pulses are usually specified in terms of speed and ratio and the standard values of these parameters are designed to match the performance of electromechanical exchange switching equipment.

1.2.1.6 *Push-button keypad*

For use with exchanges designed for dial pulse signalling there are telephones in which a numerical keypad is used to input numbers to an l.s.i. circuit which then simulates the action of a dial. The output device for signalling to line is usually a relay, typically, a mercury-wetted relay.

For electronic exchanges and some types of electromechanical exchange, a faster method of signalling known as multifrequency (MF) can be used. In this, a 12-button keypad is provided with oscillators having frequencies (see Table 9.1). Pressing any button causes a pair of frequencies to be generated, one from each band.

1.2.1.7 *Ringer (bell)*

In the idle state, the handset is on its rest, the gravity switch is open and no direct current flows in the line. The ringer is, therefore, designed to operate on alternating current. This is connected from a common supply at typically 75V r.m.s. and 25 Hz connected through a 500 ohm resistance at the exchange.

1.2.2 Types of customer (subscriber) installation

Telephones are installed at customers' (subscribers') own premises or made available to the general public at call offices (paystations) with coin collecting boxes (CCB) or card reading devices. Each of these installations is normally connected to a local exchange by a pair of wires but in isolated locations radio links may be used and optical fibres are now being used increasingly for at least the part of the connection nearest to the exchange. In some cases more than one customer's instrument may be connected to the same line which is then described as a party line. In some countries a party line, especially in remote rural areas, may serve as many as 20 customers with a system of coded ringing signals. In the UK the only party lines are those serving two customers on a system known as shared service. The case of one customer per line is known as exclusive service.

At the customer's premises there can be more than one telephone. Residential and small business users often have two or more instruments with one of several simple methods of interconnection known as extension plans. Larger businesses generally have private branch exchanges (PBX) on their premises providing connection between their extensions either under control of a switchboard operator (private manual branch exchange, PMBX) or by dialling (private automatic branch exchange, PABX).

These private exchanges are connected to their local exchange by a group of exchange lines and the public exchange selects a free line within the group (PBX hunting) when a caller dials the first line of the group.

In some cases calls incoming to the PABX are routed to an operator (attendant) who makes the connection to the required line (extension). Direct Dialling In (DDI) may also be provided so that the caller is able to dial the required extension. A PABX can also be provided as a virtual group within the public exchange, a system known as Centrex.

1.2.3 Hierarchy of exchanges

Each local public exchange provides the means for setting up calls between its own customers' lines. It is normally also connected to other local exchanges in the same area (e.g. the same town) by junctions. For calls outside this area, junctions are provided to trunk (toll) exchanges which are in turn interconnected by trunk lines, enabling calls to be established to customers in other areas. Trunk lines are also provided to one or more international exchanges which enable calls to be established to customers in the national networks of other countries.

A single exchange can sometimes combine two or more of these functions and the larger national networks have more than one level of trunk exchange in their hierarchy (e.g. three levels in the UK, four in the USA). A trunk call in the UK, therefore, passes through up to six trunk exchanges in addition to the local exchanges at each end.

In large areas, some local exchanges may be interconnected only by having calls switched through another local exchange, a technique known as tandem switching which can be independent of the trunk switching hierarchy used for calls outside the area. (See Figure 1.2.)

1.2.4 Exchange structure

The most elementary form of switching network consists of a single square switch, where connection can be made between any one inlet and any one outlet by connecting the associated horizontal and vertical circuits at the point where they cross. However, to provide an exchange capable of serving say 10,000 customers, it would not be economic either to construct a single large switch or simply to gang a large number of smaller switches to form a square matrix of 10,000 × 10,000 crosspoints. A more practical arrangement is to connect the switches in two or more stages. To illustrate this Figure 1.4 shows first a single square matrix of 81 crosspoints providing for nine paths between nine inlets and nine outlets and then how to meet the same requirement with only 54 crosspoints. The advantage of this approach is even greater with larger numbers of circuits and practical switches having typically 10 × 20 crosspoints or more, rather than only 3 × 3 as shown in the example.

This simple example provides only for the distribution of calls from the inlets to an equal number of outlets through the same number of links. In practical exchanges, it is only necessary to provide as many paths through the switching network as the number of calls which are expected to be in progress at one time. The concentration of lines on to a smaller number of paths or trunks can be achieved using the principle illustrated by another simple example in Figure 1.5.

A complete local exchange then consists of a combination of a distribution (group selection) stage and one or more line concentration units, as shown in Figure 1.6.

The separate switches are interconnected through yet another switching stage which does not provide concentration or expansion, i.e. a square switch as shown in Figure 1.4. This group switching

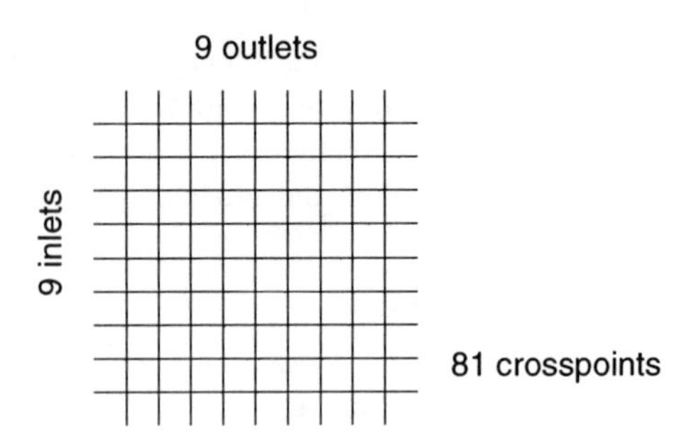

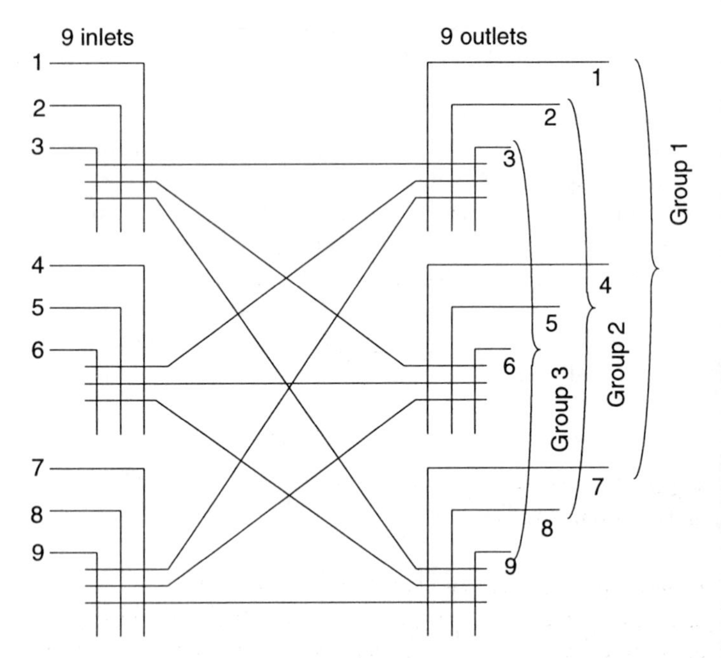

Figure 1.4 Trunking principle of distribution stage for crossbar and electronic exchanges

(distribution) stage can also be used to make connections to other exchanges. (See also Section 1.4.3.)

To complete the exchange, two other functional blocks are needed. The call connection (and disconnection) instructions have to pass

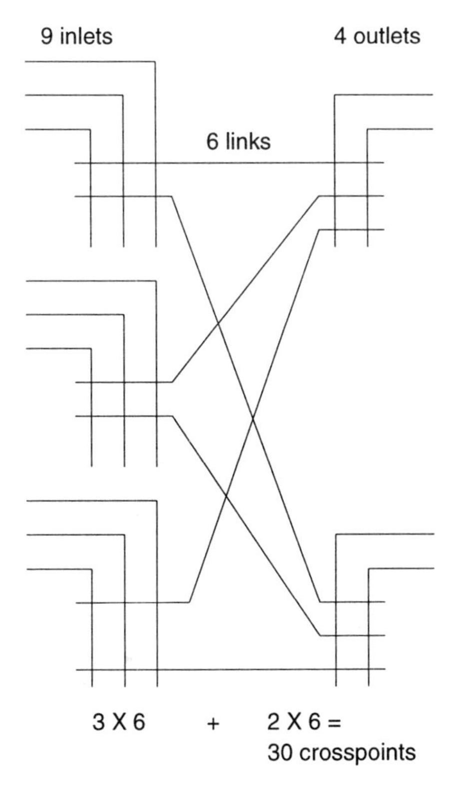

Figure 1.5 Trunking principle of concentration stage for crossbar and electronic exchanges

between the terminations and the exchange and also between exchanges for interexchange (junction) calls. These supervisory messages are conveyed by the exchange signalling systems which may transmit the relevant data over the same paths as the calls they control or they may use separate links for the purpose. Also needed is a control system for interpreting these instructions to enable appropriate paths to be set through the switches.

The operation and design of exchange switching systems can be conveniently separated into these three separate, but interdependent, functions: switching, signalling and control. In practical terms, the

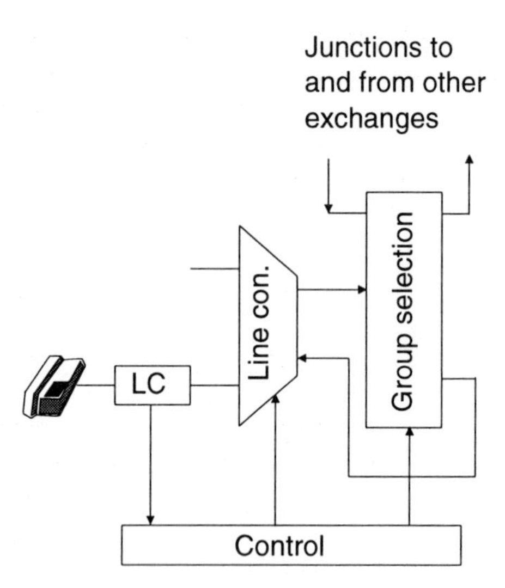

Figure 1.6 Generalised architecture of a complete crossbar or electronic exchange

choice of switching technology is very dependent upon the transmission methods used in the network concerned and this in turn influences the choice of signalling and control methods.

1.2.5 Types of exchanges

The early exchanges were all manually operated and many are still in service around the world. The exchange equipment consists of a switchboard at which an operator (telephonist) connects calls on demand.

The term 'automatic' is used to describe exchanges where the switching is carried out by machine under the remote control of the caller who could be either a customer or an operator.

The first public automatic exchange opened in 1892 at La Porte, Indiana (USA), and used rotary switches invented in 1891 by Almon B. Strowger, an undertaker in Kansas City, after whom the system is named. Since then several systems have been developed. The most successful of these employ mechanical devices called crossbar switches or arrays of reed relays. Modern versions of these systems use stored program control (SPC) in which the operations to set up each call are determined by software in the exchange processors (computers).

The new generation (digital) systems employ totally solid-state switching as well as stored program control.

1.2.6 Types of transmission

Transmission may be either analogue, in which the electrical signal represents directly the speech signal, or digital, in which speech or data are coded in pulses. At present, telephone instruments and customers' lines are generally analogue, but the trunk and junction network is increasingly digital.

The essential feature of digital transmission, as opposed to analogue, is regeneration. The digital signals consist of a sequence of digits (or symbols) each of which have a few recognised levels (typically four or less) and occupy equal periods of time known as digit (or symbol) periods. Transmission systems are designed so that corruption of the digits, due to noise, interference or distortion, does not cause the signal level in each digit period to become indeterminate. A regenerator measures the level in each digit period and generates a new digit of the original amplitude. This means that noise and corruption of the signal do not accumulate through a series of regenerators. Only two types of impairment are left. Digital errors occur when excessive noise and corruption cause a regenerator to make wrong decisions about the level and jitter, which is displacement of the digit periods from their ideal positions. Time is essentially analogue in digital transmission and so is subject to analogue types of distortion.

1.2.7 Signalling

To connect calls through a network of exchanges, it is necessary to send data referring to the call between the exchanges concerned (see Chapter 2). In the simplest case, this consists of loop/disconnect signalling similar to that used on customers' lines. Each speech circuit consists of a pair of copper wires which have direct current flowing when in use on a call and this current is interrupted to form a train of break pulses for each digit corresponding to those produced by the telephone dial. When the called party answers, the called exchange sends back an 'answer' signal to the calling exchange by reversing the direction of the current flow.

Longer circuits are unsuitable for direct current signalling e.g. due to the use of amplifiers, multiplexing equipment or radio links. Voice frequency (VF) signalling is then used in which pulses of alternating current are used at frequencies and levels compatible with circuits designed for handling speech currents (typically 600, 750, 2280 and 2400 Hz). This use of frequencies within the speech band (300-3400 Hz) is termed in-band signalling.

An alternative is the use of out-of-band (outband) signalling, in which a signalling frequency (e.g. 3825 Hz) outside the speech band is separately modulated on to the carrier in a frequency division transmission system. In digital transmission systems, the signals are digitally encoded and certain time slot(s) are reserved for this purpose separate from the speech time slot(s).

These are all channel associated signalling systems, in which the signalling is physically and permanently associated with the individual speech channel even though no signalling information is transmitted during the greater part of the call. The use of stored program control has given rise to a more efficient signalling means, known as common channel signalling. In this, one signalling channel serves a large number of speech channels and consists of a direct data link between the control processors (computers) in the exchanges concerned. Each new item of information relating to a call is sent as a digital message containing an address label identifying the circuit to which it is to be applied.

1.3 Teletraffic

1.3.1 Grade of service

It would be uneconomic to provide switching equipment in quantities sufficient for all customers to be simultaneously engaged on calls. In practice, systems are engineered to provide an acceptable service under normal peak (i.e. busy hour) conditions. For this purpose, a measure called the grade of service (GoS) is defined as the proportion of call attempts made in the busy hour which fail to mature due to the equipment concerned being already engaged on other calls. A typical value would be 0.005 (one lost call in 200) for a single switching stage. To avoid the possibility of serious deterioration of service under sudden abnormal traffic it is also usual to specify grades of service to be met under selected overload conditions.

1.3.2 Traffic units

Traffic is measured in terms of a unit of traffic intensity called the Erlang (formerly known as the traffic unit or TU) which may be defined as the number of call hours per hour (usually, but not necessarily, the busy hour), or call seconds per second, etc. (Mazda, 1996). This is a dimensionless unit which expresses the rate of flow of calls and, for a group of circuits, it is numerically equal to the average number of simultaneous calls. It also equates on a single circuit to the proportion of time for which that circuit is engaged and consequently to the probability of finding that circuit engaged (i.e. the grade of service on that circuit). The traffic on one circuit can, of course, never be greater than one Erlang. Typically a single selector or junction circuit would carry about 0.6 Erlang and customers' lines vary from about 0.05 or less for residential lines to 0.5 Erlang or more for some business lines.

It follows from the definition that C calls of average duration t seconds occurring in a period of T seconds constitute a traffic of A Erlangs given by Equation 1.1.

$$A = C\frac{t}{T} \tag{1.1}$$

An alternative unit of traffic sometimes used is the c.c.s. (call cent second) defined as hundreds of call seconds per hour (36 c.c.s. = 1 Erlang).

1.3.3 Erlang's full availability formula

Full availability means that all inlets to a switching stage have access to all the outlets of that stage. Trunking arrangements, in which some of the outlets are accessible from only some of the inlets, are said to have limited availability.

For a full availability group of N circuits offered a traffic of A Erlangs, the grade of service B is given by Equation 1.2.

$$B = \frac{\frac{A\,!}{N}}{1 + A + \frac{A^2}{2\,!} + \ldots + \frac{A^N}{N\,!}} \tag{1.2}$$

This formula assumes pure chance traffic and that all calls originating when all trunks are busy are lost and have zero duration.

1.3.4 Busy hour call attempts

An important parameter in the design of stored program control systems, in particular, and common control systems in general, is the number of attempts to be processed, usually expressed in busy hour call attempts (BHCA). Typically a 10,000 line local exchange might require a processing capacity of up to 80,000 BHCA.

1.3.5 Blocking

If two free trunks cannot be connected together because all suitable links are already engaged the call is said to be blocked. In Figure 1.4, if one call has already been set up from inlet 1 to outlet 1, an attempted call from inlet 2 to outlet 2 will be blocked. The effects of the blocking are reduced here by allocation of the outlets to the three routes.

Blocking can be reduced by connecting certain outlets permanently to the inlet side of the network to permit a blocked call to seek a new path. Alternatively, a third stage of switching may be provided.

It is possible to design a network having an odd number of switching stages in which blocking never occurs. Non-blocking arrays need more crosspoints than acceptable blocking networks and are uneconomic for most commercial telephone switching applications.

1.3.6 Network routing

Telephone networks are normally based on a hierachical arrangement and the higher levels are often fully interconnected, e.g. the UK transit network. Dimensioning is simplified by the use of fixed routing but the overall Grade of Service is sensitive to faults and overloads on individual links. All routes must, therefore, be dimensioned to overload conditions.

It is possible to use Automatic Alternative Routing based on overflow, although the dimensioning is more difficult since the overflow traffic is not random and, therefore, Erlangs loss formula does not apply.

If unlimited overflow is allowed, there is a danger of affecting the performance of traffic which normally uses the secondary route.

A method sometimes used in SPC systems is known as Trunk Reservation, whereby the overflowing traffic is only given access to the secondary route if more than a set level of circuits are free, while the traffic normally using this route always has access to it.

1.3.7 Queuing

A queuing system provides facilities for storing calls in order until the server is available. The number of calls which can be held in a queue at any one time must be limited and, therefore, dimensioning must take account of the loss probability as well as delay conditions.

Many practical systems involve multiple queues dependent on the service required and this results effectively in a queuing network which can be quite complex to analyse.

1.4 Analogue switching

1.4.1 The manual exchange

The first telephone exchanges were operated by hand. Instructions were passed verbally between the subscriber and the operator (telephonist). Similar principles were used on early telegraph exchanges. As the network grew and the operators were expected to handle increasing volumes of traffic, the signalling and control elements were developed, which formed the basis of the first automatic systems.

In its more common form, the manual switchboard has subscribers' (customers') terminations connected to jacks (sockets). The "links" consist of a pair of plugs on flexible cords (wires) which can be used to interconnect any pair of jacks. Signalling consists of calling indicators (e.g. relays and lamps) associated with the jacks and clearing indicators associated with the cords and plugs. Call routing instructions are passed verbally.

The simplest type is the magneto board on which the caller gains the attention of the operator by using a hand generator to send an a.c. signal which is detected by an electromechanical indicator at the switchboard. An improved design is the central battery (CB) board which uses gravity switch contacts at the telephone to operate a relay at the exchange to light a lamp at the switchboard. In both cases the caller then tells the operator verbally what call to make.

The great strength of the system is that the 'control' is human and, therefore, intelligent and the variety of connections that can be made is virtually unlimited. Services such as advice of duration and charge, transfer of calls when absent, wake up calls etc., which are so complex to provide automatically, present no problem at all on manual exchanges.

The weakness of the manual exchange, which has led to its almost complete disappearance, was essentially its slowness. Unless operators could set up calls faster it would have been impossible to find enough operators to handle the volume of traffic on the modern network.

1.4.2 Step by step

Ratchet driven switches, remotely controlled directly by pulses from the calling telephone, are used to simulate the operator's actions on a plug and cord switchboard. The wipers (moving contacts) are connected by short flexible cords and perform a similar function to the switchboard plugs. The place of the jack field is taken by a bank of fixed contacts and the wipers are moved one step from each contact to the next for each pulse by means of a ratchet and pawl mechanism.

In early Strowger exchanges, each customer was provided with a 100-outlet two-motion selector. The expense of providing one large switch or selector is now avoided by using a stage of smaller cheaper uniselectors to connect the callers to the selectors as and when they require them. Expansion of the exchange beyond 100 lines is by adding further stages of selectors.

The setting up of a call progresses through the exchange stage by stage. Each selector connects the call to a free selector in the next stage in time for the train of pulses to set that selector. The wipers step vertically to the required level under dial pulse control at 10 pulses per second then rotate automatically at 33 steps per second to select a free outlet by examining the potential on each contact. This is step-by-step operation and, when used with Strowger selectors, it needs only simple control circuits which can be provided economically at each selector.

Concentration is provided at the uniselectors. The final two digits of the number are used to position the final selector (connector) wipers to the contacts connected to the required line, thus providing the expansion function.

The final selectors also contain the circuit element known as the 'transmission bridge' which provides current to the calling and called telephones from the central supply (battery) at the exchange. It is here that 'answer' and 'clear' conditions are detected to control charging (metering) and release of the connection.

A large city often has a large number of exchanges with overlapping service areas where it would be unacceptable for callers to have to consult a different list of dialling codes according to which exchange the telephone they happen to be using is connected. To allow the same list of codes to be used anywhere in the area, director equipment is provided which translates the code dialled by the caller into the actual routing digits required.

Even with the addition of the director, the step-by-step system lacks flexibility. The convention of ten finger holes in a dial, means that the switching network branches in a decadic fashion. Each switch requires a decadic ability and is mechanically designed accordingly. If, however, traffic is unevenly distributed throughout the exchange numbering range, some parts of this capability will be under-used. The relative slowness of the switches also makes it impractical to attempt rerouting for call transfer or to by-pass congestion in the network. The mechanical complexity of the switches themselves presents reliability problems and requires expensive maintenance.

Even on these systems there are some calls which customers either cannot or will not set up for themselves. To provide for these, numbers are allocated (e.g. 100) which can be dialled for assistance. Such calls are connected through the automatic exchange to a switchboard where the operator can establish the required connection also through the automatic equipment. These switchboards are called automanual exchanges.

1.4.3 Register control (Crossbar)

In crossbar and analogue electronic systems, the equipment operates at high speed. It functions on an end-to-end instead of the step-by-step basis used by Strowger exchanges and the trunking pattern does not conform with the decimal notation used by the telephone dial. For these reasons, it is necessary to interpose some equipment during the setting up process which will convert the customers' dialling signals into information which can be used by the switching equipment. When a call is originated, therefore, the control sets crosspoints to connect the calling line to a register which has functions broadly similar to those of an operator. The register detects and stores the dial (or MF key) pulses and signals them to the control. The control, which is duplicated or sectioned in some way for security, see Section 1.7, analyses the digits received and sets the appropriate crosspoints to establish the call. Being common, the control equipment can be used to set up only one connection at a time but operates so quickly that this restriction is not noticeable to customers.

The director, as used in the Strowger system, is an example of register control. Some of the disadvantages of the Strowger switches have been overcome with other designs of mechanical switch but they are generally unsuitable for direct control from the customer's dial. Register control provides a way of using such switches.

The most successful register controlled systems are based on the crossbar switch which is mechanically simpler than the rotary switches of the Strowger system and can be designed for greater reliability and smaller size. Translation facilities like those provided by the director system, but not limited by considerations of mechanical switch sizes, are readily obtained because there is no inherent relationship between number allocation and particular switch outlets.

The Crossbar switch mechanism consists of a rectangular array of contact sets or crosspoints, which can be selected by energising the electromagnets corresponding to their vertical and horizontal coordinates. Typical mechanisms contain 200 or more crosspoints per switch. The switches are generally used in combinations to form a group selection (distribution) stage and a combined concentration/expansion stage, as shown in Figure 1.7. Additional crossbar switches may be used for establishing other connections necessary for the operation of the exchange e.g. RF, which connects registers, as required, to callers via the concentration stage CS.

The function of the register is to receive the caller's instructions in the form of dialled digits and convert them into the appropriate signals for route selection. Operation of the relevant magnets in the required crossbar switches in the concentration stage (CS) and the group selection stage (GS) is carried out by the markers.

Each marker is directly associated with a set of crossbar switches which it serves exclusively. The registers are available in a common pool and are assigned as required to incoming calls.

The subscriber's line circuit (LC) detects the calling condition and signals to the relevant marker which causes CS to connect the calling line to a free transmission bridge (TB). This in turn cooperates with

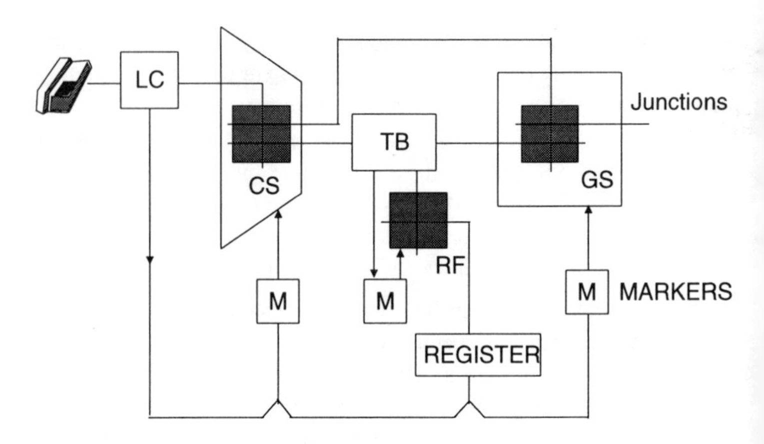

Figure 1.7 Crossbar exchange

another marker to set a connection through the Register Finder (RF) to seize a free register.

Unlike the Strowger system, the number of switching stages need not be related to the exchange numbering scheme.

1.4.4 Central control (Reed systems)

The use of registers represents one application of 'common control' where one control circuit (e.g. a marker or a register) is shared by several switches.

Greater centralisation of the control, however, offers some further operational advantages. When numbering changes are made during the life of the exchange, for example, it may be possible to avoid the need to change a large number of circuits such as registers. It is also possible to employ more effective search patterns to find free paths, thus making economies possible in switch provision. For example, in register systems, the register chosen initially for the call does not necessarily have an unrestricted choice of switch paths for the forward connection of the call. It may have access only to some of the switches at the next stage and it cannot govern the choice of free switches beyond that. Although suitable free paths may exist the exchange may be unable to allocate them and the customer may fail to get the call.

In any block of switches under one control, it is only possible to set one call at a time. To extend the control across a whole exchange needs faster switches and a control speed to match. Although some central control has been used with crossbar switches, it has been more widely applied to reed relay exchanges in the public network and later to fully electronic exchanges.

The reed relays consist of sealed contact units inside an operating coil. They are interconnected to form switch matrices, similar to Figure 1.1, in which each crosspoint has its own reed relay. Being sealed and having no moving parts, they are more reliable than crossbar switches.

A typical arrangement for a reed relay exchange with central control is shown in Figure 1.8. This is the system known as TXE4 which is further described in Chapter 3.

This system shows a return to the concepts of the simple manual exchange. All terminations on the exchange are connected to the same set of switches and the complete operation of setting up a path is controlled by a processor, which corresponds to an operator and can observe directly the state of the whole exchange. Several processors are provided per exchange and they work independently of one another, as did the operators on a multiple position switchboard,

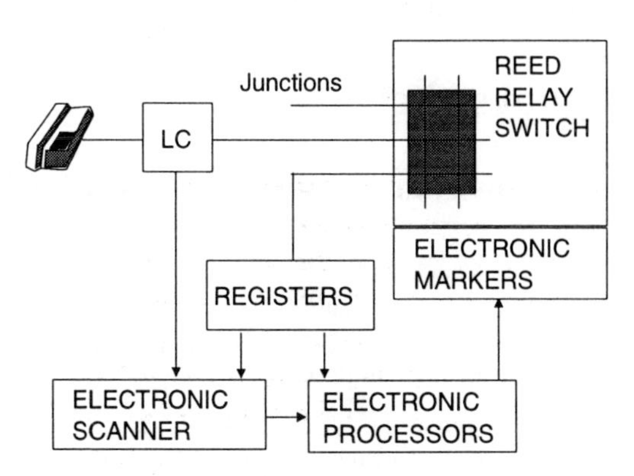

Figure 1.8 Reed electronic exchange

sharing only common directory information and operating rules. Provided the processors are made powerful enough, the systems organised in this way have again begun to acquire the power and flexibility associated with an operator-controlled manual exchange.

1.4.5 Distributed control

In the Strowger system, control elements are associated with each element of the switch so that control is distributed with no centralisation. Communication between control elements is minimal, instructions being received directly from the subscriber at the relatively slow rate of ten pulses per second.

Although centralisation of the control brought many benefits it also has its problems. Exchanges grow to meet traffic demand during their 'life' and their processing power requirements grow with them. This is not easy to achieve economically if all the processing power resides in one central processor. Also, it is usual to provide at least two central processors to guard against failure yet doubling the capacity in this way is wasteful.

Various architectures have, therefore, been devised which retain the essential functional centralisation of processing while allowing decentralisation of the processors themselves. This facilitates growth and permits a more economical replication strategy.

1.5 Digital switching

1.5.1 Space division and time division switching

Analogue exchanges such as Strowger, crossbar and reed relay systems employ space division switching in which a separate physical path is provided for each call and is held continuously for the exclusive use of that call for its whole duration.

Digital and other pulse modulated systems employ time division switching in which each call is provided with a path only during its allocated time slot in a continuous cycle. Each switching element or crosspoint is shared by several simultaneous calls, each occupying a single time slot, in which a sample of the speech is transmitted. This sample may be a single pulse as in pulse amplitude modulated systems or a group of pulses as in pulse code modulated (PCM) systems.

Speech itself is analogue and so too are most telephones. Analogue to digital (A/D) conversion (modulation) is, therefore, necessary at some point in each connection if digital switching is to be employed. For customers connected to existing analogue local exchanges this occurs when the call first encounters a PCM transmission trunk on leaving the exchange. For customers connected to a digital local exchange it is necessary to provide the A/D conversion on each line. The cost of this conversion constitutes a major part of the exchange cost but is amply compensated for by the overall economy of digital switching and the performance advantages in the transmission network.

The European standard is for speech encoded in 8 bit bytes at 64kbit/s. The US standard is 7 bit coding at 56kbit/s.

1.5.2 Digital tandem (transit) exchange

This interconnects transmission systems which are multiplexed in groups usually of 24 or 30 speech channels on each link. The tandem exchange has to be able to connect an incoming channel in one PCM link to an outgoing channel in another PCM link. Received speech samples must, therefore, be switched in space to the appropriate link and translated in time to the required channel time slot.

A simple time switch is shown in Figure 1.9, which provides for time slot interchange between the 30 channels of an incoming PCM circuit and the 30 channels of an outgoing circuit.

Each successive byte corresponding to a channel is written cyclically into the appropriate location in a 30 × 8 bit RAM under the control of a time slot counter synchronised with the incoming circuit clock. The contents of the store are read out to the outgoing circuit under the control of a connection store. The words in the store, which is itself addressed cyclically, identify which address in the speech store is to be addressed. If the incoming time slot counter and connection store run in synchronism, the outgoing circuit carries a

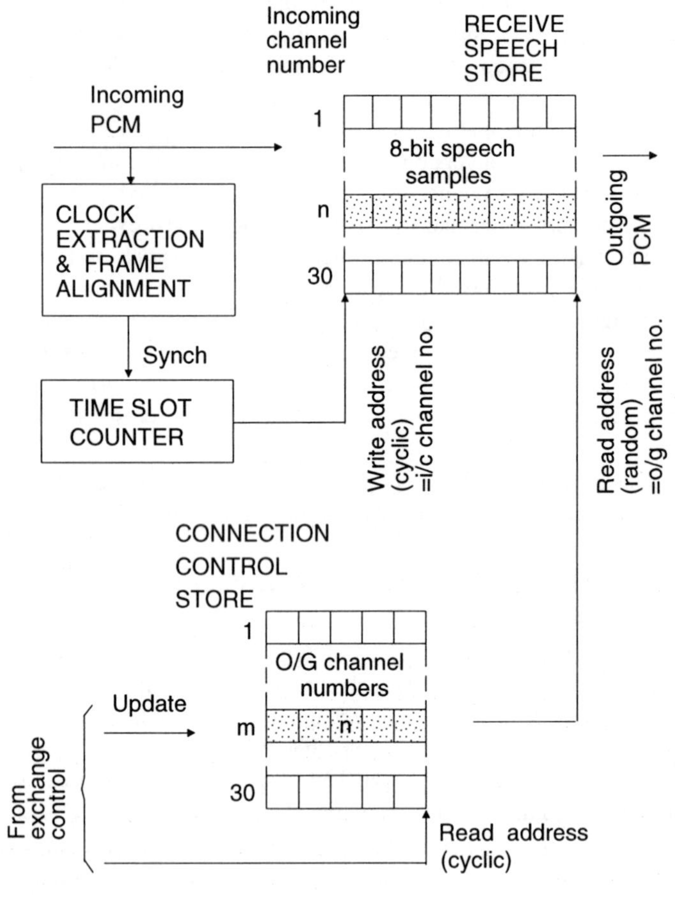

Figure 1.9 Simple time switch

similar multiplex to the incoming circuit but with the timeslots in a different order as determined by the contents of the connection store. In practice, the connection store may have a different speed or a different number of channels (words) from the incoming circuit.

Practical PCM systems employ additional channels for signalling and framing purposes. These can be extracted for use by the exchange control by providing additional RAM addresses from which they are read out on to a circuit connected to the exchange control processor. Each incoming circuit, therefore, requires 32×8 bit RAM addresses.

For large exchanges, each time switch serves a group of say 32 incoming circuits using 1024 RAM addresses. The connection store then runs at 32 times the incoming multiplex data rate to interleave all 30×32 speech channels on a single high speed highway to the next switching stage in the exchange.

A simple TDM space switch is shown in Figure 1.10 consisting of a matrix in which the crosspoints are electronic gates controlled by a connection store. Each incoming timeslot on each incoming circuit is switched individually to the appropriate outgoing circuit, in accordance with the addresses stored in the space switch connection store.

A complete tandem exchange might use a time switch, like Figure 1.9, to connect each incoming circuit to a row in a matrix, like Figure 1.10. The verticals of the matrix could be the outgoing circuits but are more usually highways connected to a time switch associated with each outgoing trunk group. This arrangement is shown in Figure 1.11 and is known as time-space-time (TST) switching.

If the internal exchange bit rate is high enough to interleave all the incoming channels on to a single highway, no space switch is required. Each cross-office timeslot corresponds to a unique outgoing circuit/timeslot combination so no outgoing time switch is required either. Allocation of highway timeslots to outgoing channels is achieved by gating on a fixed pattern of strobe clock pulses. This is sometimes known as TS switching, the output gating being regarded as a 'space' (S) switching stage. It is also sometimes called memory switching. The size of exchange is limited by the available technology since RAM size and speed requirements are proportional to the number of channels provided across the exchange. The continuous evolution of integrated circuit technology makes this technique increasingly attractive. Larger exchanges can be constructed by interconnecting separate TS stages.

One other possible configuration is space-time-space (STS). This was used for some early designs when the only available time switching technology was to insert expensive time delay elements in the speech path.

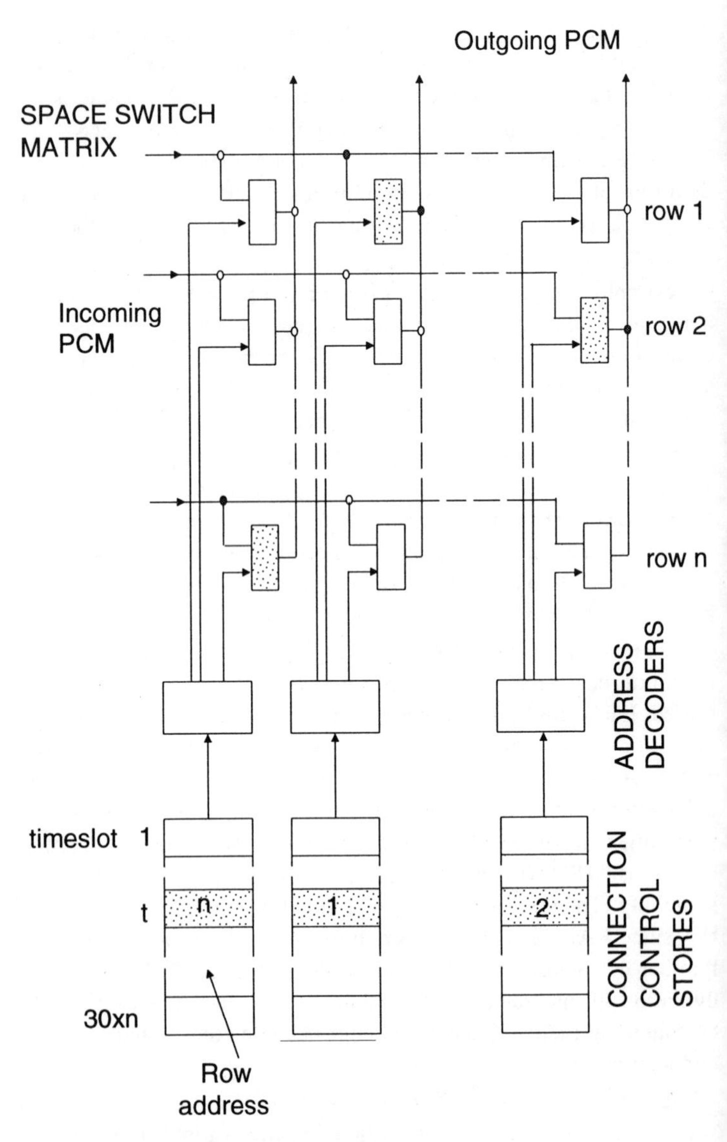

Figure 1.10 Simple space switch

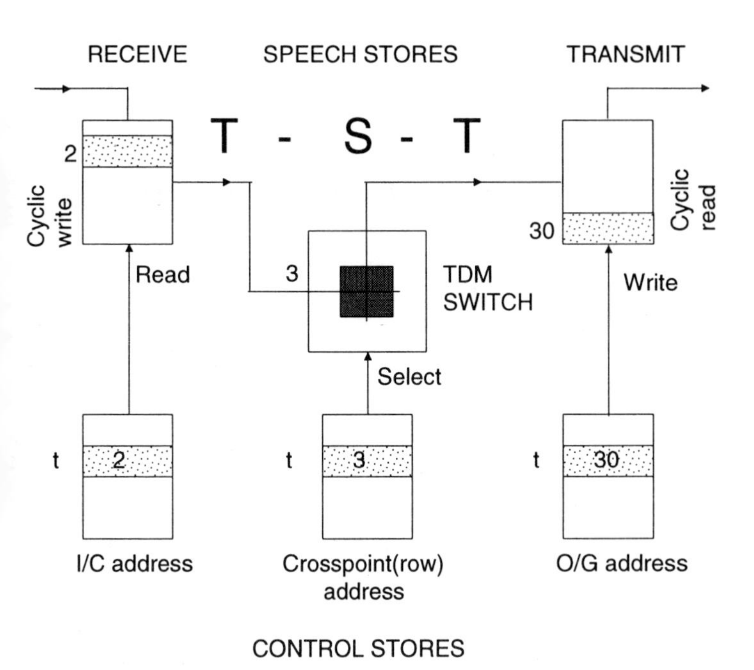

Figure 1.11 T - S - T switch

The arrangements described above carry speech in one direction only. All 'receive' channels are connected to one side of the switch, regardless of whether the circuits are incoming or outgoing in the traffic sense and all 'transmit' channels are connected to the other side. Each call requires transmission in each direction and these two paths can occupy the same crosspoint in the space switch but at different time slots, usually 180 degrees out of phase. Separate time switches must be provided in each direction of transmission.

1.5.3 Digital local exchange

The switching networks in Section 1.5.2 provide only for the distribution of calls from the inlets to an equal number of outlets through the same number of links. In practical exchanges, it is only necessary to provide as many paths through the switching network as the number of calls which are expected to be in progress at one time. (See also Section 1.3.)

The concentration of lines on to a smaller number of paths or trunks is provided in the form of an additional time switching stage. The number of available time slots to the main switch is less than the total number of input time slots and store locations in the concentrator speech stores they serve. This might consist typically of up to 16 groups of 256 lines, each written cyclically into a 256 word store but with only 256 timeslots available for allocation to the acyclic read out from all these stores.

Telephone customers' lines are generally on individual pairs employing analogue transmission. The interface to the analogue customer's line includes certain functions which require voltage and power levels incompatible with the digital concentrator switch and which, therefore, have to be provided at the individual line circuit (line card). The line interface functions are generally referred to as the Borscht functions:

Battery feed to line.
Overvoltage protection.
Ringing current injection and ring trip detection.
Supervision (on-hook/off-hook detection).
Codec for analogue to digital conversion.
Hybrid for two-wire to four-wire conversion.
Test access to line and associated line circuit.

The connection between the concentration stage and the main switching stage is a PCM multiplex and very suitable for connection over an external transmission line. This is not true of analogue exchanges where the connection consists of a pair of wires per link (channel).

Digital exchanges are, therefore, particularly suited to provision as a single large central switch serving a number of remote concentrators.

1.6 Stored program control

1.6.1 Processors

The functions of a digital exchange switch, described above, need processor control to interpret the connections required in terms of signals received and connection store contents to be maintained.

One advantage of central control by processors, is that facility changes may be effected by changes to a few processors instead of many registers. This advantage is further increased if the processors are based on 'software' programs. This method of operation is known as stored program control (SPC) and the exchange shown in Figure 1.8 has a large part of its logic contained in stored program. It is possible, in the limit, to convert into processor software almost all of the logical functions of the exchange, including signalling.

The advantages are that the variety of hardware may be minimised, so easing production and stocking problems, and that facility changes may be readily accomplished. Disadvantages are that software development can be exceedingly complex and that real time problems may occur.

1.6.2 Software

The essence of SPC is that it is software driven. The functionality of the exchange control processors can be changed to suit different hardware configurations (number and type of lines, traffic etc.) for different installations and growth during the life of the system. The features offered can also develop as requirements change over time (e.g. tariff structures, dialling codes) without major hardware replacements.

This software must, of course, meet the same standards of reliability and availability as those demanded of the switch and processor hardware. Software does not wear out like hardware. Failures can only occur due to design faults (bugs) which have not been detected during the design and development phase. The number of combinations of external inputs and system states in programs of this size makes it impossible to guarantee the complete absence of such bugs.

To provide some degree of fault tolerance, it is first necessary to design the software in discrete modules with defined interfaces between them. This enables more comprehensive testing of possible combinations within each module than would be possible on the complete system. It also allows checks to be applied to input and output conditions at the interfaces. These checks can either be used to trigger the hardware changeover mechanism or to reinitialise the software module to an acceptable state.

The modular design of software also facilitates the evolution of features during the life of the system and the implementation of distributed control.

The primary function of the software is to set up and clear down calls on demand. Calls arise concurrently in real time and in a random manner. In addition, the software is required to carry out diagnostic checks on the hardware, provide statistical data on traffic and performance and provide call records for billing purposes. Separate modules are provided for each of the functions in the applications software. The operating system schedules the use of the hardware resources by these application programs, storing the status of all registers etc. when a program is deactivated and restoring the same state when it is reactivated. This multiprogramming technique enables high priority tasks (e.g. call set up) to be given precedence over lower priority tasks (e.g. statistics).

Figure 1.12 shows in simplified form a typical modular break down of the software in an SPC exchange. The line scan program continuously updates a tabular store of the state of each line termination (e.g. free, busy, calling etc.) and alerts the central call routing software to connection requests. The latter validates the request against a pre-determined class of service for the termination and receives called number data from the line scanner or the signalling software, depending on the type of termination. The received digits are compared with the stored routing data (translation table) to enable the switch to be controlled via the connection control stores (Figures 1.9 to 1.11). Exchange staff can interrogate the status of the processor

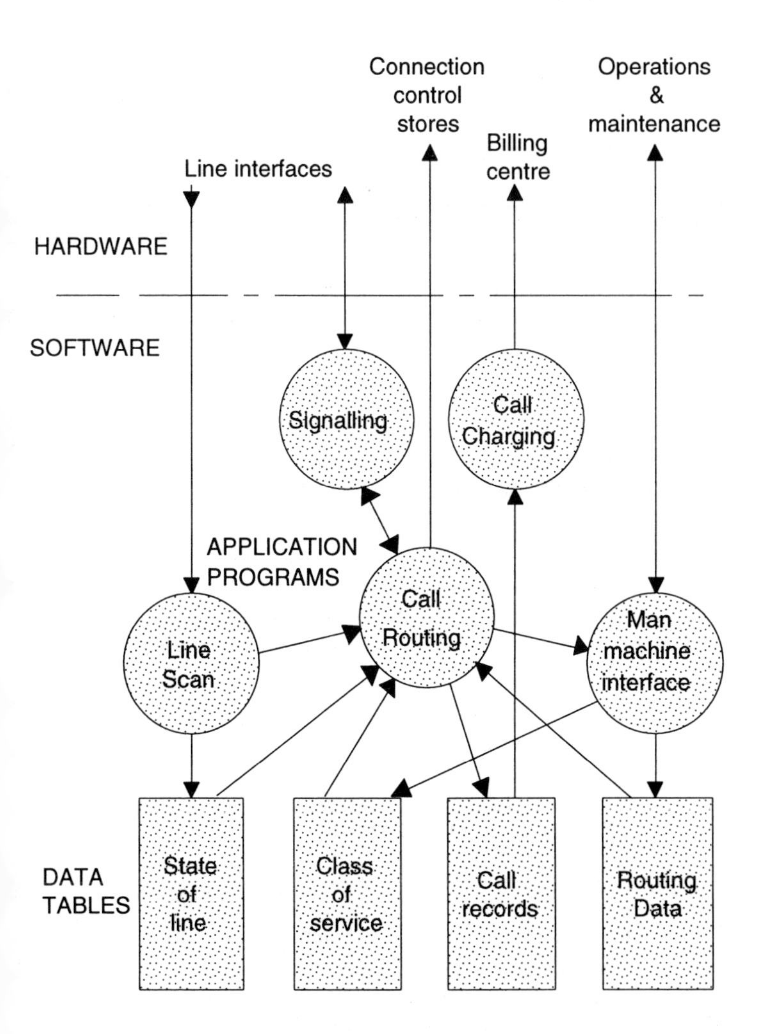

Figure 1.12 SPC software structure

and update the routing and class of service tables through the Man Machine Interface (MMI).

1.6.3 Network aspects

The introduction of SPC has enabled electronic exchanges to provide customer and administration features not readily available with earlier control systems. Examples include the ability to transfer calls at will and allow calls to wait on a busy line. Where new features are self contained within an exchange they can be introduced by installing a new build of software. If the operation of a feature requires identical operation at all exchanges, the practical problems of modifying them all at once become immense.

One solution is to download new programs, routing data etc. remotely via data links and to site the operations and maintenance terminals at a central location to serve a group of exchanges. Central access to the databases and processors of all the exchanges permits short term as well as long term changes to be made to facilities and routing. This is known as Network Management and enables plant failures to be identified, located and traffic to be re-routed around congestion bottlenecks. A further development of the centralised control principle is the concept of the Intelligent Network in which individual exchange processors interrogate a remote database for instructions on how to process particular types of call, or route such calls to a central service provider.

1.7 Operational security

Unlike many other systems which can plan for regular downtime for maintenance, telephone switching systems are required to provide virtually uninterrupted service 24 hours a day, seven days a week, year in – year out, to tens of thousands of customers (and junctions). At the same time they must be adaptable to growth and operational requirements. Occasional faults are, of course, inevitable and the system design must allow for this, so that a fault does not result in loss of service to any significant number of customers. It is better for a fault to temporarily reduce the grade of service to all customers rather than to deny service entirely to a smaller number. As a result, various system architectures have been evolved.

In electromechanical systems, each call follows a path wholly exclusive to itself, sufficient paths being provided to carry the predicted traffic. Failure in any path, therefore, reduces the grade of service to all terminations normally having access to it, but since other paths are available, service is not wholly cut off to any one. Electromechanical systems are, therefore, inherently fault-tolerant as regards path switching and, for example, Strowger type selectors incorporate their own local control which reacts to loop-disconnect signalling over the path concerned.

In electronic (common control) systems, security is achieved by building in redundancy and by the use of inbuilt check circuits and fault detection which will both route calls round a faulty condition and also indicate the presence of the fault so that maintenance action can correct it.

In the switching area, security is normally provided by giving any termination access to multiple paths through the network. A particular example of this is the use of sectionalisation where the switching network is divided into planes of switching, operating in parallel and all accessible to all terminations. Thus a fault affecting any one plane of switching will not cause loss of service to any terminations, the traffic being carried by the remaining planes, albeit at a slightly reduced grade of service.

In the control area, security is normally provided by replication. In its simplest form this would mean a single duplicated control unit. The two units can operate:

1. In parallel with a fault cutting out the defective unit.
2. As worker and spare with automatic changeover in the event of a fault.
3. Alternatively under the control of a time cycle.

However, except in very small systems, a single duplicated control is unlikely to be adequate and a number of alternative options are available:

1. Triplication, where three identical units are used in parallel with majority decision equipment on the outputs.

2. One-in-N Sparing, where a number of similar equipments are involved and one (or more) spares are provided such that, in the event of a fault, one of the spares may be switched into service in place of the faulty equipment.
3. Load sharing, where a number of similar equipments operate in parallel each carrying a proportion of the traffic. In the event of a fault, the equipment concerned is taken out of service with the traffic being carried by the remainder at a reduced grade of service.

In practice, some combination of these options is often used with the overall control area being sub-divided dependent on the function to be performed.

In the case of traffic dependent equipment such as registers, these are normally dimensioned to carry the predicted busy hour traffic at the required grade of service in the presence of one or more faults, depending on the predicted fault rate.

Whichever method is employed to provide the operational security, it is essential that adequate check circuits and routines are provided to detect the faults together with appropriate means for reporting them (e.g. lamp display, audible alarm or print out) such that the fault can be remedied within the mean time between failures (MTBF) for the equipment concerned.

1.8 References

Atkinson, J. (1950) *Telephony*, Pitman.

Bear, D. (1988) *Principles of Telecommunication Traffic Engineering*, Peter Peregrinus.

Berkeley, G.S. (1949) *Traffic and Trunking Principles in Automatic Telephony*, Benn.

Briley, B.E. (1983) *Telephone Switching*, Addison Wesley.

Hughes, C.J. (1986) Switching — the state-of- the-art, *Br Telecom Technol J*, Part 1, January; Part 2, April.

Hills, M.T. and Kano, S. (1976) *Programming electronic switching systems*, Peter Peregrinus.

Hughes, C.J. (1986) Switching — State of the art, *British Telecom Tech.*, **4**, (1) and (2).
Mazda, F. (1996) *Analytical techniques in telecommunications*, Butterworth-Heinemann.
Redmill, F.J. and Valdar, A.R. (1990) *SPC digital telephone exchanges*, Peter Peregrinus.
Smith, S.F. (1978) *Telephony and Telegraphy*, 3rd edn, Oxford University Press.
Takamura, S. et. al. (1979) *Software design for electronic switching systems*, Peter Peregrinus.
Vickers, R. and Wernik, M. (1988) Evolution of switch architecture and technology, *Telecommunications*, May.

2. Signalling systems

2.1 Analogue network signalling

The basic signalling on subscriber lines comprises the following:

1. Supervisory, the on-hook, off-hook conditions of the subscribers.
2. Address information from the caller which may be decadic dialled pulses or push-button signals.

These signalling functions are transferred over the switched network by signalling systems. Call processing can be performed by the above two functions. A further function 'operational' is necessary to enable the best use to be made of the network and to cater for system and operating organisation facilities.

The type of switching equipment, direct control or common control, and the type of transmission equipment have significant influence on the type of signalling system.

2.1.1 Type of switching system

Direct control switching systems, e.g. step-by-step, do not allow for separation of the supervisory and address signalling functions. The dialled address positions the switches directly, the numbering scheme controlling the routing of call connections. The supervisory and address signalling functions are combined in the network line signalling systems. The signalling limit tends to be limited by the permissible decadic address pulse distortion.

In common control switching (Figure 2.1), the common control equipment (register, translator, marker) deals with the call connection set-up through the switchblock, being released from a call on connection establishment to deal with other call connections. The marker

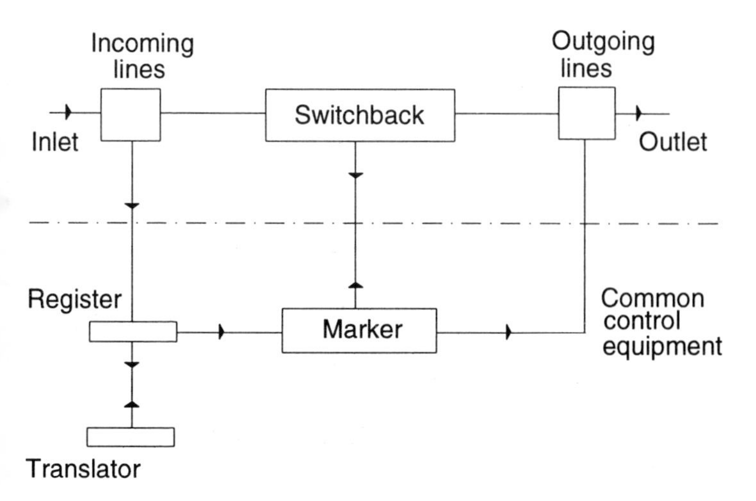

Figure 2.1 Generalised principle of common control switching

controls the switching. The incoming line is the incoming mark at the switchblock. Address information inputs the register which refers to the translator. The resulting translation establishes the outgoing mark identifying the outgoing line. The marker switches the two marks together through the switchblock. The translation facility permits automatic alternative routing.

Common control switching permits separation of the supervisory (line) and address (selection) signalling functions over the network (Figure 2.2).

2.1.2 Influence of type of transmission system

Analogue network transmission plant varies in type: audio, FDM, point-to-point PCM. As analogue network signalling is on the call speech path, the type of transmission plant has influence on the signalling method.

A.C. signalling within the voice range (300Hz to 3400Hz) can be applied to any type of transmission media, but this application rationalisation is not economic since supervisory line signalling systems are

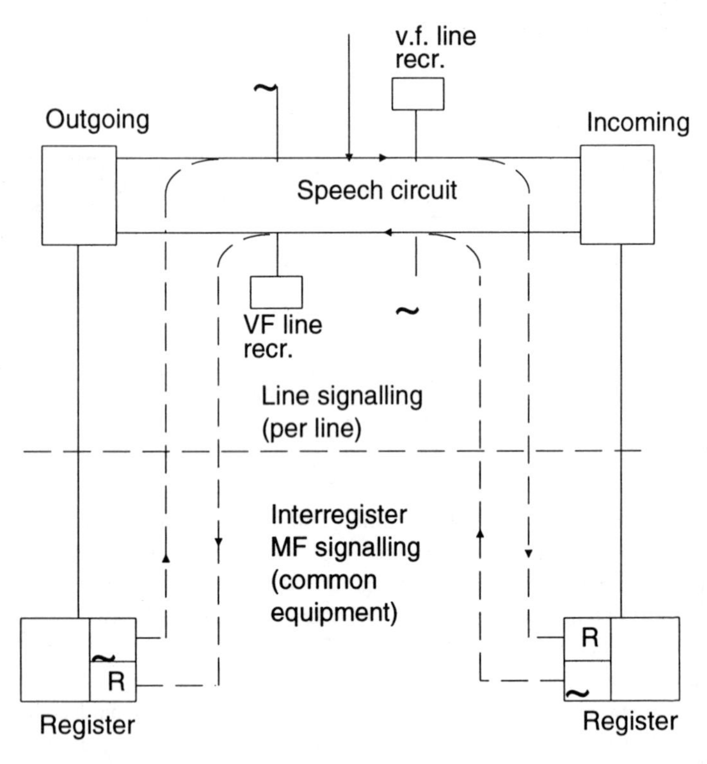

Figure 2.2 Line and interregister signalling

equipped per line. Line signalling is of type (d.c., VF etc.) according to the type of line transmission.

A connection switched through an analogue exchange is supervised. A supervisory unit is equipped per line switched for this purpose. The unit incorporates a transmission bridge and detects supervisory signal conditions on the calling and called sides independently. Line units associated with external lines enable electrically independent signal conditions to be used in the exchanges and those on the external lines. The supervisory unit is then often combined with such line units.

2.1.3 D.C. line signalling

2.1.3.1 *Short-haul d.c. line signalling systems*

There are a number of different forms (Welch, 1979): loop disconnect d.c., loop reverse battery, single wire d.c., high/low resistance d.c., battery and earth pulsing. Loop disconnect d.c. signalling on 2-wire lines is the simplest and most widely used.

The basic signal repertoire is:

1. Forward; seizure, address information (when direct control switching), clear forward.
2. Backward; answer, clear back.

Figure 2.3 shows the principle of loop disconnect d.c. line signalling. Its operation is as follows:

1. Seizure. Relay A at A operates to the caller's off-hook loop.
2. Address. (Assuming direct control switching.) Relay A at A responds to the caller's dialled pulses. When switched through exchange A, contact A1 loops the line to operate relay A at B. A1 loop disconnects the line, so transmitting the remaining dialled pulses to relay A at B.
3. Answer. Relay D at B operates to the called subscriber's off-hook loop. D1 and D2 reverse the line d.c. polarity to operate the rectified relay D at A (or a polarised relay as in the loop reverse battery d.c. system).
4. Clear back. The called subscriber's on-hook disconnect releases relay D at B, contacts D1 and D2 restoring releases relay D at A.
5. Clear forward. The caller's on-hook disconnect releases relay A at A. A1 releases relay A at B (assuming the more usual calling party release).

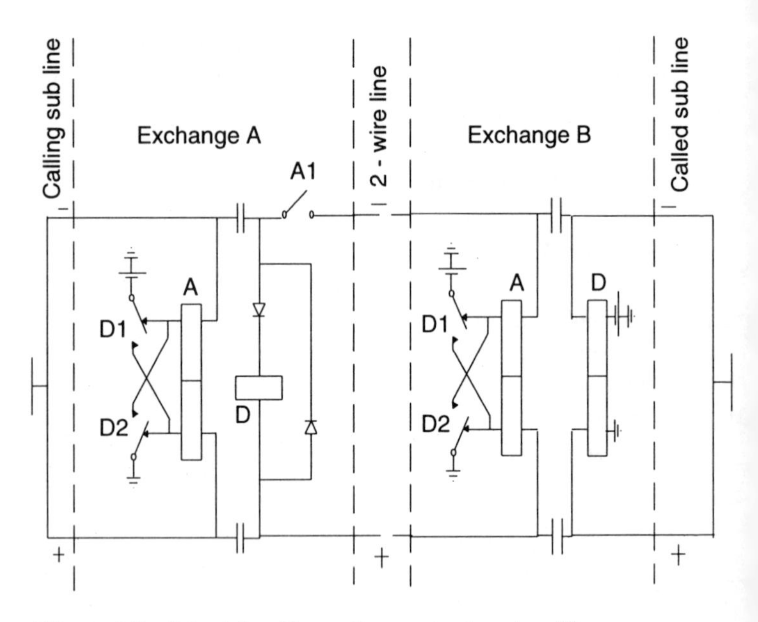

Figure 2.3 Principle of loop disconnect d.c. signalling

With common control switching, address signalling is inter-register and would not be conveyed by the line signalling system.

2.1.3.2 *Long distance d.c. line signalling (LDDC)*

LDDC is applied on lines beyond the limits of short-haul d.c. signalling when a metallic circuit is available (Welch, 1979). The phantom of the 4-wire circuit is used for signalling. The method extends the signalling limit by the use of sensitive polarised relays and the decadic address pulsing limit by a more efficient pulsing technique than loop disconnect.

With loop disconnect pulsing, the send end impedance varies during the decadic pulsing process (substantially zero on pulse make and infinite on pulse break). This variation results in asymmetric waveforms and thus to decadic pulse distortion on repetition. Further, the single current pulsing receive relay, with fixed operate and release current values, gives varying pulse distortion when operating on asymmetric waveforms.

LDDC necessitates symmetrical pulsing waveforms. To achieve this:

1. The send end impedance is equal during the pulse make and break periods.
2. The pulsing battery is at the send end (it is at the receive end in short-haul).
3. The polarised receive relay operates in one direction for the make, and in the other for the break pulses.

The above requires the provision of both outgoing and incoming line signalling terminals. Both terminals incorporate the transmission bridge.

There are different designs of LDDC systems. The use of polarised relays for signalling sensitivity and for bi-directional operation to minimise decadic address pulsing distortion is a common principle.

2.1.4 A.C. line signalling

2.1.4.1 *Low frequency a.c. signalling*

This uses a below the speech band signalling frequency. Signalling imitation by speech does not arise. Typical frequencies are 25, 50, 80, 135, 150 or 200Hz. D.C. signals are converted to a.c. for transmission, these being converted to d.c. at the distant end. The system may be applied to metallic circuits above the range of short-haul d.c., and to circuits isolated for d.c. when d.c. signalling is not possible. The more usual application is on 2-wire circuits, the signals usually being pulsed since only one transmission path is available. The technique has declined in importance.

2.1.4.2 *Voice frequency (VF) line signalling*

The signalling frequency is inband, within the speech range of 300Hz to 3400Hz of f.d.m. transmission. The system is completely flexible in that it can operate on any circuit affording speech transmission. The signals are amplified by transmission system amplifiers. 4-wire operation is the more usual, the forward and backward signalling paths being separate, the duplex signalling being achieved by two simplex signalling paths (Figure 2.4).

In 4-wire operation, the VF receivers are permanently associated with the respective speech paths. As speech may contain the signalling frequency, the system is subject to signal imitation by speech and must incorporate features to safeguard against this.

A buffer amplifier of unit gain forward and some 60dB loss in the reverse direction protects the signal receiver from near-end interference from the switching equipment. The signalling is usually link-by-link on multi-link connections and an arrangement of line splits adopted to isolate the VF signal to the particular link. Signalling during speech is inadmissible.

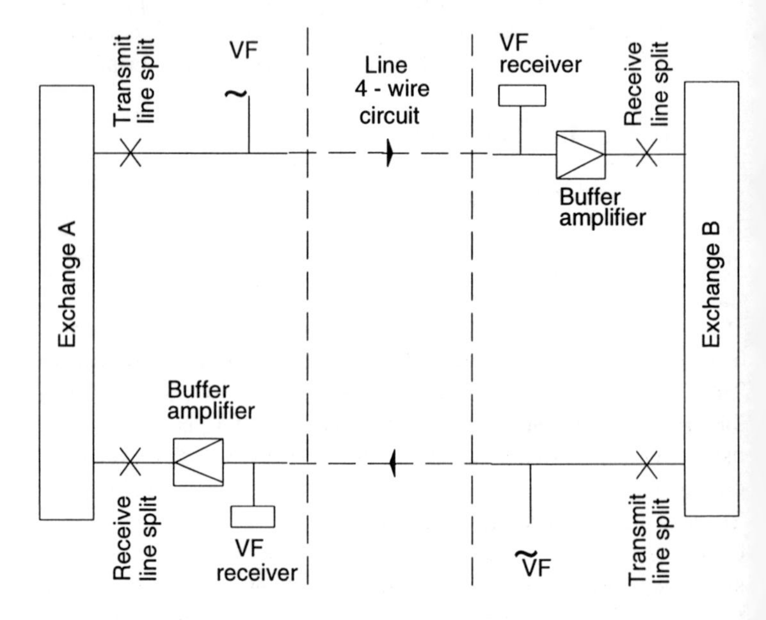

Figure 2.4 General arrangement of VF line signalling system

Protection against signal imitation (false line splits or false signalling) relies on exploiting differences between speech and signal currents, e.g.:

1. Adoption of a signal frequency which is less liable to exist, or persist, and has less energy than other frequencies in speech. This indicates a reasonably high signal frequency, typically, U.K. 2280Hz, North America 2600Hz.
2. Speech currents containing the signal frequency usually have other frequencies. The signal receiver has two elements: (a) tuned to the signal frequency, and (b) responsive to other frequencies to oppose receiver operation to (a).
3. Signals made longer than the normal persistence of the signal frequency occurring in speech. At least 40ms longer for important signals such as release.
4. Signals of two frequencies compounded are less liable to occur, and persist, in speech than one frequency. 2 VF signalling systems are known, but 1 VF systems are more usual.

The VF signals may be continuous or pulse. Pulse signals of different lengths permit a greater signal repertoire. A continuous signal requires a continuous response to cease it.

2.1.4.3 *Outband signalling*

This is a.c. signalling within a 4kHz channel of FDM transmission. The channel bandwidth is divided into a speech channel (300Hz to 3400Hz) and a signalling channel by filtration. One frequency only is used, located approximately midway between two adjacent 4kHz channels.

The forward and backward signalling paths are separate on the 4-wire circuit, the duplex signalling being achieved by the two simplex signalling paths. Signalling frequency 3700, 3825, or 3850Hz is typical; the ITU-T recommends 3825Hz. E-lead and M-lead control applies for outband signalling.

Unlike inband VF signalling, outband signalling:

1. Is not subject to signal imitation by speech or other interference.
2. Eliminates the line splits.
3. Permits signalling during speech (e.g. meter pulse transmission).

On the other hand VF signalling can be applied to any type of transmission media, whereas outband signalling can only be applied to FDM channel circuits.

As outband signalling is independent of speech, there is virtually complete freedom in the choice of signalling mode with simple arrangements. The signalling may be:

1. 2-state continuous tone-on idle (off during speech).
2. Tone-off idle (on during speech), which however tends to overload the transmission system.
3. Semi-continuous, tone-off idle and off during speech (not preferred).
4. Pulse.

With the 2-state continuous signalling mode in each direction, outband signalling enables the basic loop d.c. signalling condition to be simulated far more easily compared with inband VF signalling, making for much simpler signalling terminals. For the general application, tone-on idle (off during speech) is preferred.

2.1.5 PCM signalling

Applied when a PCM system is applied point-to-point in an analogue network. The ITU-T specifies for two PCM systems: 30-channel (32 time slots) and 24-channel (24 time slots).

2.1.5.1 *30-channel PCM*

Time slot 16 is used for signalling for the 30 speech channels in each direction. A 16-frame multiframe technique gives 16 (0 to 15) appearances of time slot 16. The 8 bits of time slot 16 of frames 1 to 15 of the multiframe are divided into two groups of 4 bits each. Thus, in every 2ms (the multiframe time) 4 out-slot signalling bits are available for each speech channel. The coding gives 15 signal possibilities for each speech channel (code pattern 0000 is not used). This is ample for line signalling requirement. Should inband interregister MF signalling apply, the PCM system deals with this as for speech.

The logic behind the approach was that time slot 16 would be the 64kbit/s bearer for common channel signalling in the future integrated digital networks. The PCM system would be retained in service, but with signalling change.

2.1.5.2 *24-channel PCM*

This is the North American system. In the pioneer D1 version, bit 8 of the octet in each time slot is used for in-slot signalling, the speech bits being reduced to 7. This only gives two signal conditions (0000 --- and 1111 ---) in each direction, a very limited repertoire.

The later D2 version adopts a 12-frame multiframe. Bit 8 of the time slots in every sixth frame (called signalling frames) is used for signalling, giving two in-slot signalling bits per time slot. This slows the signalling relative to D1, reduces the speech bits in the signalling frame time slots to 7, but increases the signal repertoire.

2.1.5.3 *Interregister signalling*

Applied with common control switching, the signalling terminals being associated with the registers (CCITT, 1988a; Miller, 1971). The system deals primarily with address signalling and is separate from line signalling systems (Figure 2.2).

The address signalling may be decadic d.c. pulsing, but is more usually 2-out-of-6 multifrequency (2/6mf), giving potential for 15 signals in each direction (Figure 2.5). The signalling is non-decadic, the 2/6mf combination giving the address digit value. This contributes to fast signalling. On 2-wire lines, the two simplex signalling paths are obtained by filtered bandwidth separation.

Subscriber and system facilities often require network signalling. The increased signal repertoire of 2/6mf compared with decadic

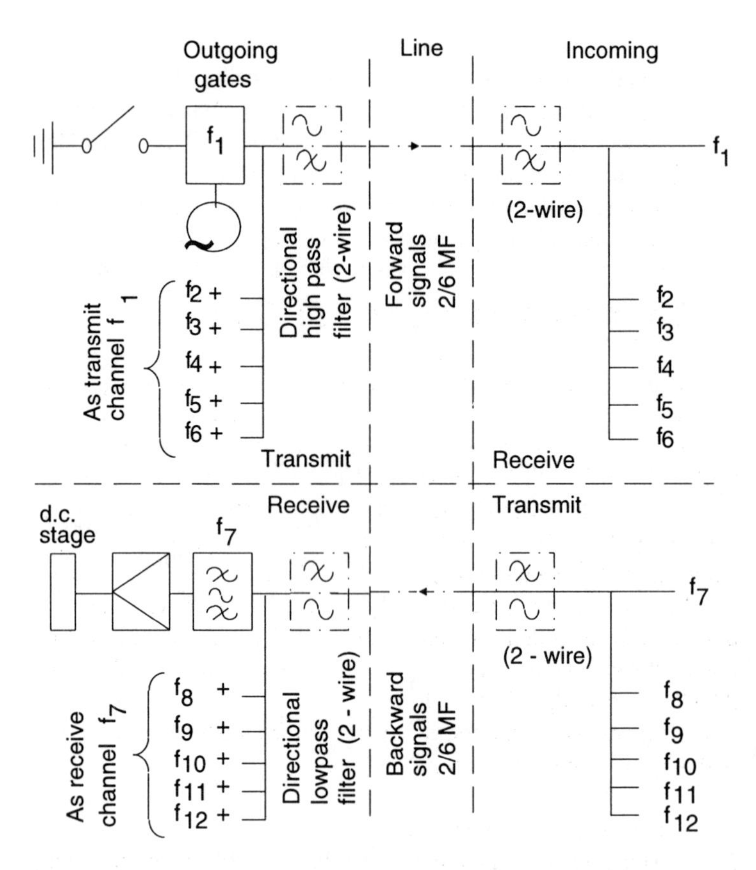

Figure 2.5 General arrangement of interregister IF signalling system

interregister signalling enables the network to be exploited for facilities.

The 2/6mf signals may be pulse or continuous (compelled). In compelled, a continuous acknowledgement backward signal ceases the forward information signal, which cessation ceases the acknowledgement.

A number of different detail 2/6mf signalling systems exist, but are the same in principle. The ITU-T R2 system is based on:

1. Continuous compelled signalling.
2. Forward signals 120Hz spaced in the band 1380Hz to 1980Hz, backward signals 120Hz spaced in the band 1140Hz to 540Hz.
3. To increase the signal repertoire if required, both the forward and backward 2/6 frequency combinations have primary and secondary meanings by a shift condition.

2.2 Common channel signalling

Stored programme control (SPC) of switching prompted a reappraisal of the network signalling technique. Processor control makes possible the concentration of signalling logic for a large number of traffic circuits. With SPC it is inefficient for the processor, which works in the digital mode, to deal with on-traffic-circuit analogue signalling. A much more efficient way of transferring information between SPC exchanges is to provide a bi-directional high speed data link between the two processors over which they transfer signals in digital form by coded bit fields. A group of traffic circuits (many hundreds) would thus share a common channel signalling (CCS) link. All CCS signalling is link-by-link.

CCS gives rise to the following requirements which do not arise with signalling on traffic circuit systems:

1. High order error rate performance.
2. Signalling backup to maintain service on failure or unacceptable performance.
3. Assurance of traffic circuit continuity as, unlike on-traffic-circuit signalling, CCS does not establish traffic path integrity.
4. Circuit label included in the signal messages to give traffic circuit identity.

CCS may be associated or quasi-associated signalling. In associated (Figure 2.6) the CCS link is between the same locations as the traffic circuits served.

In Figure 2.7, associated signalling applies AC and CB. The signalling for traffic circuits AB is routed ACB (and BCA) in the quasi-associated mode, C being a signal transfer point. Should asso-

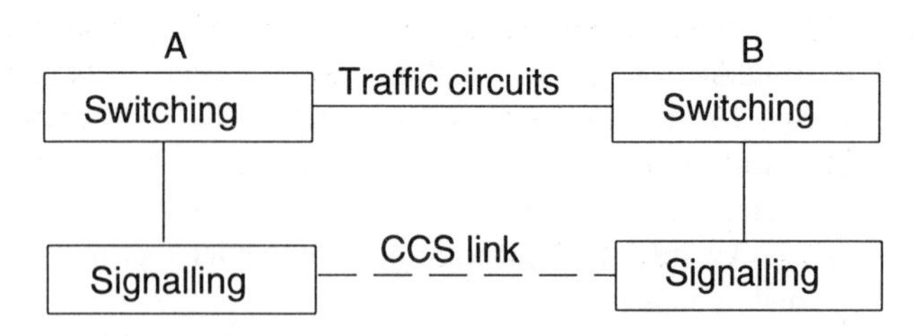

Figure 2.6 Example of associated CCS between A and B

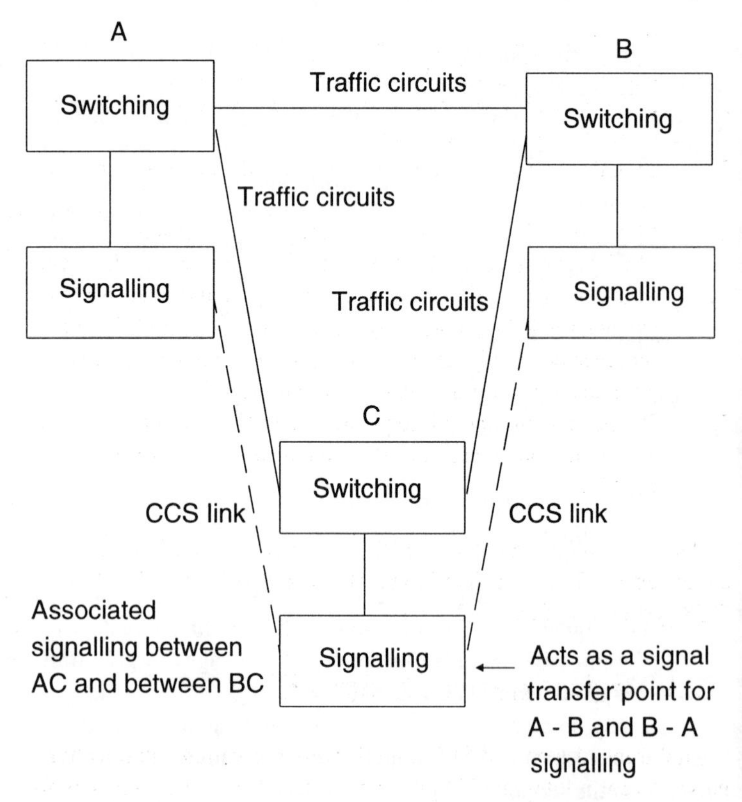

Figure 2.7 Example of quasi-associated signalling between A and B and between B and A

ciated CCS be equipped AB, on failure of any one of the associated signalling links AB, AC or BC, backup quasi-associated signalling ACB, ABC or BAC may apply to maintain service; C, B or A acting as signal transfer points.

2.2.1 ITU-T No.6 CCS system

This was the pioneer CCS system (Welch, 1979; CCITT, 1988; Dahlbom, 1972). Its main features are:

1. Signalling bit rates 2.4kbit/s (analogue link), 4kbit/s and 56kbit/s (digital link). Other rates as required.
2. Error control; error detection by redundant coding and error correction by retransmission.
3 Signal units (SUs) 28 bits each of which the last 8 bits are error check bits.
4. SUs grouped into blocks of 12, the 12th and last SU of each block being an acknowledgement, i.e. the number of the block in the other direction being acknowledged and the error control acknowledgements, positive that the relevant SU has been received error free; negative indicating error detected and thus a request for a re-transmission of that SU.
5. Blocks are numbered sequentially, but individual message SUs in the block are not, being identified by their position in the block.

The later requirement for ISDN indicated the need for a more advanced CCS system than No.6. This resulted in the ITU-T No.7 CCS system (see Section 2.3.1.)

2.3 Digital network signalling

Digital transmission and SPC digital switching systems in telephone networks form Integrated Digital Networks (IDN). The logical evolution of IDN is to Integrated Services Digital Networks (ISDN). This is achieved by extending the IDN to the user by providing Integrated Digital Access (IDA). Present ISDN is narrowband (N-ISDN), based on the 64kbit/s channel capacity of the digital networks. Figure 2.8 identifies the signalling requirements for N-ISDN.

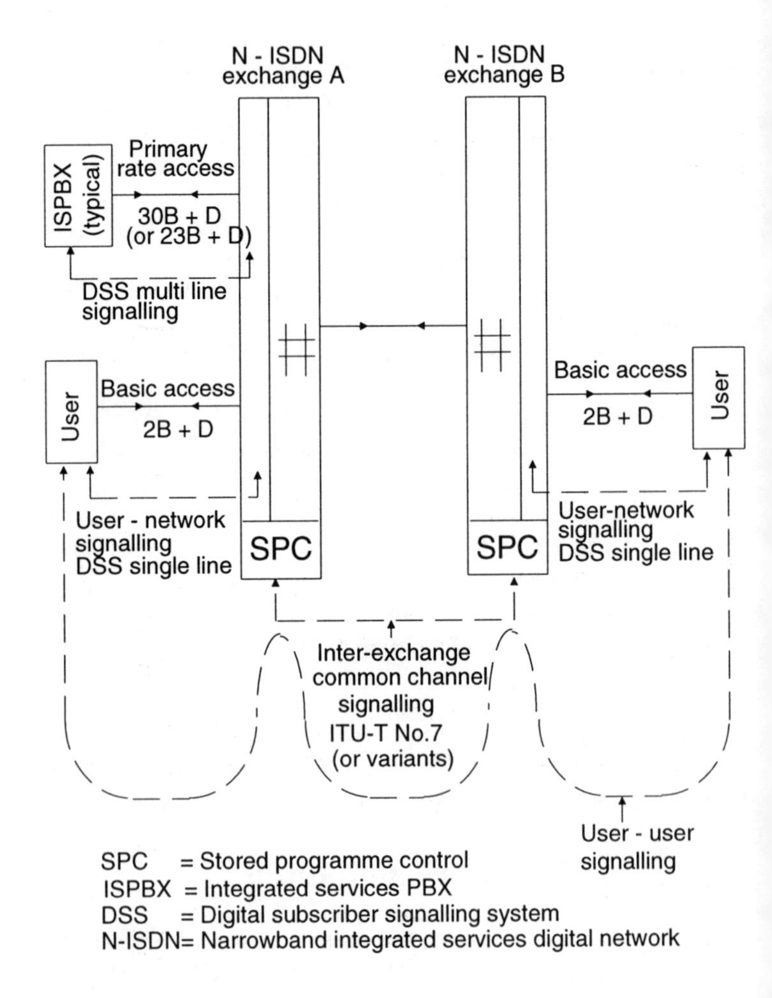

Figure 2.8 N-ISDN signalling requirements

2.3.1 ITU-T No.7 inter-exchange CCS system

The pioneer ITU-T No.6 CCS system was produced primarily for the analogue environment, the digital versions being produced by subsequent study, but retained the basic features of the analogue system (Welch, 1979). In the result, No.6 has limitations for ISDN. The No.7 system is optimised for the ISDN. The ITU-T No.7 international specification is framed as a common vehicle, admitting variants for national application. The signalling bit rate is 64kbit/s time slot 16 30-channel PCM. Figure 2.9 shows the No.7 general arrangement.

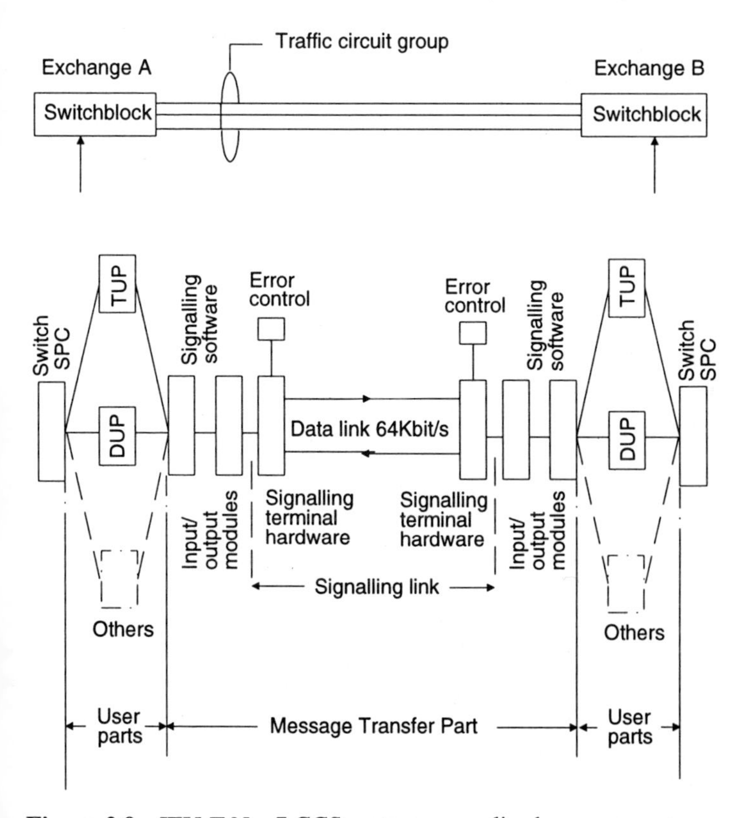

Figure 2.9 ITU-T No. 7 CCS system generalised arrangement

2.3.1.1 *User parts*

No.7 adopts a functional concept (Figure 2.9), allowing change to a function without significant impact on others (CCITT, 1988c). User parts (telephony TUP, data DUP, etc.) are service dependent parts of the signalling system controlling message requirements for particular services. This allows system evolution for additional services by additional User Parts, permitting the SPC processor to be a common utility from the signalling aspect. Appropriate coding of a bit field in the signal messages indicate the service.

2.3.1.2 *No.7 formatting principles*

The signal messages are HDLC octet byte variable bit length signal units (SUs) with opening and closing flag delimiters coded 01111110 (CCITT, 1988d). The circuit label includes the originating and destination exchanges point codes, plus the circuit identification of the traffic circuit directly connecting the originating and destination points.

H0 codings identify the various signal groups (address signals, call supervision signals, etc.). H1 codings identify particular signal messages in a relevant H0 group; e.g. answer signal (H1 coding) in the call supervision group (H0 coding).

2.3.1.3 *No.7 error control*

The basic error control is error detection by redundant coding (CRC) and error correction by re-transmission (CCITT, 1988e). This requires positive and negative acknowledgement signals to be returned from the receive end. The former clears the signal message from the retransmission store at the transmit end on correct receipt; the latter causes the signal message and all other signal messages which may be in the retransmission store to be retransmitted. The signal messages are sequence numbered.

Idle (fill-in) SUs are transmitted when no message SUs are available for transmission. Idle SUs, not deposited in the re-transmission store and not retransmitted on error detected, are CRC error checked; this being a form of preventative maintenance.

A variant of the above basic error control is adopted for long propagation time signalling links (e.g. satellite). In this case the whole content of the retransmission store is retransmitted repeatedly when there are no new signal messages to be sent (so called preventive cyclic retransmission). This is a form of forward error correction.

Receive end monitors measure the error rate, automatic change-over to the backup signalling facility being initiated on unacceptable performance. The backup may be associated or quasi-associated signalling. Figure 2.10 shows a typical associated signalling backup. The up-to-four signalling links, but at least two, are signalling load shared. Assuming a four link signalling module, in normal operation

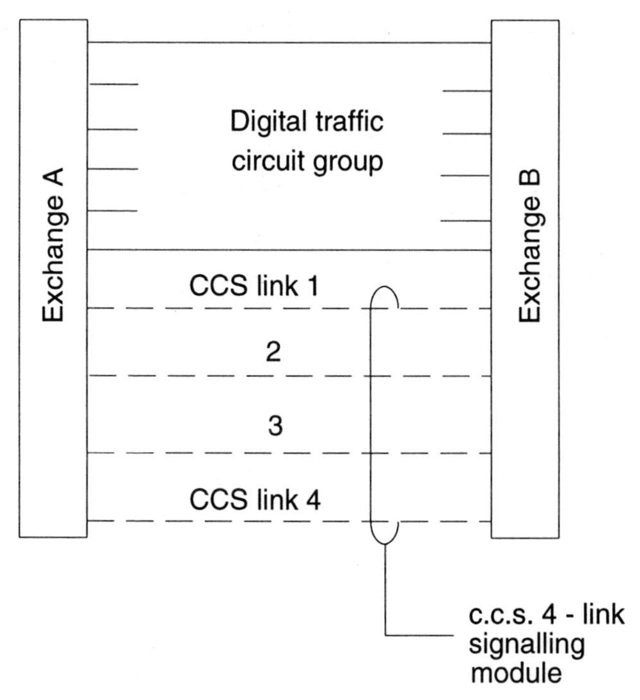

Figure 2.10 Common channel signalling module principle

each signalling link serves, typically, 1000 traffic circuits. In extreme error conditions the one remaining CCS link carries the total signalling load for 4000 traffic circuits.

The polynomial for the error check is given by expression 2.1. The CRC (sometimes called FCS, frame check sequence) is a 16 bit field.

$$x^{16} + x^{12} + x^{5} + 1 \tag{2.1}$$

2.3.2 User-network signalling

Digital Subscriber Signalling No.1 (DSS 1 is the ITU-T name for the N-ISDN access protocols.

In N-ISDN, the network is extended down to the user's location, terminating in a Network Termination (NT). The NT forms the interface between the user's Terminal Equipment (TE) signalling and the network signalling and performs the signalling conversion. The ITU-T regulates for a concept of S and T reference points at the interface (CCITT, 1988f).

The main types of user-network capabilities recommended by the ITU-T are (CCITT, 1988g):

1. Basic access 2B + D.
2. Primary rate access 30B + D (30-channel PCM networks) or 23B + D (24-channel PCM networks).

B is a 64kbit/s traffic channel; D the signalling channel (16kbit/s in basic access, 64kbit/s in primary rate access).

2.3.2.1 *Basic access signalling*

Information TE-NT and NT-TE is transferred in 48 bit 250 microsecond frames in each direction (CCITT, 1988h; CCITT, 1988i). Each frame contains four traffic octets (two B1 and 2 B2), four outslot D signalling bits giving the 16kbit/s signalling channel, plus other information. 4-wire operation applies on a passive bus.

The signalling in the two directions NT-ISDN exchange, which may be on a 2-wire user access line, adopts the HDLC protocol with opening and closing flag delimiters and error check by a CRC bit field.

A command and response technique is recommended, a command signal message being discarded if checked as error. The command signal is transmitted continuously within a certain time period and ceased by receipt of an error free response. CCITT, 1988j gives the detail of the various signal messages.

2.3.2.2 *Primary rate access signalling*

Here, the ISPEX (or LAN etc.) incorporates the NT function, the access being 4-wire to the N-ISDN exchange.

The D signalling channel is 64kbit/s time slot 16 in 30-channel PCM (time slot 24 in 24-channel PCM). Common channel signalling is adopted, the system serving the 30 B traffic channels (23 B in 24-channel PCM). The signal messages, CRC checked, are in sequence numbered HDLC frames delimited by opening and closing flags.

Frames may be retransmitted continuously until an error free acknowledgement of correct reception is received from the other end of the access link.

CCITT, 1988k, gives the detail of the various signal messages.

2.3.3 User-to-user signalling

This facility may be used to exchange information between two N-ISDN users, the information transferred being, for example, data or keypad/display. The user-to-user signalling may be performed:

1. Over the circuit switched B traffic path connection between the two users.
2. Over an end-to-end signalling connection between the two users (Figure 2.8).

The user-to-user information is carried transparently, and is not interpreted, by the network.

The network is informed by the calling user that the user-to-user signalling facility is a requirement. This requirement is included in the call set-up message from the caller.

2.4 References

Buatois, E. (1990) A platform for No. 7 signalling, *Telecommunications,* August

CCITT (1988a) *Specifications of signalling system R2*, Blue Book Vol.VI Fasc. VI.4, Recomms. Q.350 – Q.368, ITU, Geneva.

CCITT (1988b) *Specifications of signalling system No.6*, Blue Book Vol.VI, Fasc. VI.3, Recomms. Q.251 – Q.300, ITU, Geneva.

CCITT (1988c) *Introduction to CCITT signalling system No.7*, Blue Book Vol. VI. Fasc. VI.7, Recomms. Q.700, ITU, Geneva.

CCITT (1988d) *No.7 functions and codes of the signal unit fields*, Recomms. Q.703, item 2.3.

CCITT (1988e) *No.7 error control*, Recomms. Q.703 items 4-6.

CCITT (1988f) *ISDN user-network interface — reference configurations*, Blue Book, Vol.111 Fasc. 111.8, Recomms. I. 411, ITU, Geneva.

CCITT (1988g) *ISDN user-network interfaces, — interface structures and access capabilities*, Recomms. I. 412.

CCITT (1988h) *Basic user-network interface*, Recomms. I. 430.

CCITT (1988i) *Primary rate user-network interface*, Recomms. I. 431.

CCITT (1988j) *Digital access signalling system, data link layer*, Blue Book, Vol. VI Fasc. VI.10, Recomms. Q921, ITU, Geneva.

CCITT (1988k) *Digital access signalling system, network layer*, Recomms. Q931.

Dahlbom, C.A. (1972) Common channel signalling, — a new flexible interoffice signalling technique, *IEEE International Switching Symposium Record*, Boston, U.S.A.

Flaminio, H. (1992) Synchronising SS7, *TE&M*, January 1

Hardisty, A. (1991) Progress in private network signalling, *Communications International*, September

Leber, M.J. (1991) Performance needs in the CCS 7 signalling network, *Telecommunications*, August.

Miller, C.B. and Murray, W.J. (1971) Trunk transit network signalling systems — multifrequency interregister signalling, *Post Off. Electr. Eng. J.*, **63** Pt.1. 43-48.

Welch, S. (1979) *Signalling in Telecommunications Networks*, IEE Telecommunications Series 6, Peter Peregrinus Ltd., London.

Wittering, S. (1994) QSIG: Signalling the future, *Communications International*, December.

Zoicas, A. (1990) DTMF needs no extras, *Communications International, April.*

3. Telephone exchanges

3.1 Introduction

Chapter 1 introduced the principles of switching systems. There are currently many switches being used throughout the world. A large number of these are analogue, although they are gradually being converted over to digital switching systems.

In this chapter some of the switches in current use are described. It is not the intention to provide a comprehensive description of these switches, but rather to cover their architecture, and show how they implement the basic switching functions. Because of space limitations only a few of the many major switches are described. These are the TXE analogue switch from STC, now part of Nortel; the System 12 switch, originally from ITT but now being developed by Alcatel Bell; the AXE switch from Ericsson; and the DMS switch from Nortel.

3.2 TXE analogue exchanges

The TXE2 and TXE4 analogue exchanges are used in the U.K. for small and large exchanges respectively. These systems employ a switching matrix having a similar structure to crossbar systems but using reed relays for the crosspoints. These have the advantage of being sealed against corrosion and dirt and operate in only about 2ms, making them more amenable to electronic control than Strowger or crossbar switches. They also have a separate operating coil for each crosspoint enabling switch sizes to be optimised for system requirements.

The higher speed of operation and the use of electronic control permits greater use of self checking features to improve reliability and quality of service. The greater use of electronics enables these systems to benefit from advancing technology in that field, in terms of size and versatility in the provision of customer facilities.

3.3 The TXE2 system

TXE2 is designed for use as a small local exchange. Its capacity is limited basically by the number of busy hour calls it can handle. The theoretical maximum capacity of a TXE2 unit is 240 erlangs of total traffic (i.e. originating, terminating and tandem) and the multiple is determined by the total traffic per line; typically a single unit of TXE2 can serve from two to three thousand lines. The lower limit is basically an economic one and could be about 200 lines in circumstances where the growth rate is fairly large.

Normally a TXE2 will be installed as a single unit within the stated capacities for lines and traffic. Should either of these grow beyond the capacity of the single unit, a second unit may be coupled to the first to double the exchange capacity. In exceptional circumstances, growth beyond the capacity of two exchange units could be met by using unit switching equipment to combine three or more exchange units.

The TXE2 system may be divided into three main areas:

1. The switching area comprising three stages A, B and C in the outward direction and four stages D, C, B and A in the inward direction.
2. The supervisory and register area comprising outgoing junction, incoming junction and local supervisory relay sets, together with a common pool of registers serving all three types.
3. The control area comprising calling number identification, called number marker and call control equipment.

3.3.1 The switching area

The switching network is a general purpose network as shown in simplified form in Figure 3.1. The sizes of the various switches can be varied to meet differing requirements of lines and traffic but are

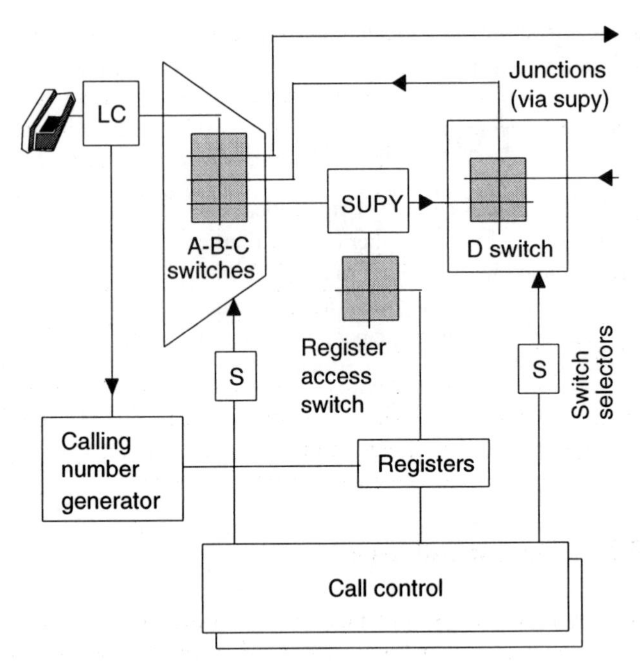

Figure 3.1 TXE2 system. (Compare with Figure 1.7)

multiples of a standard switching module comprising 25 crosspoints arranged in a 5 × 5 reed relay matrix. The A switch serves to concentrate customers' lines onto B switch inlets and the number may be varied to ensure optimum loading. The B and C switches are interconnected to obtain full accessibility between every B switch inlet and every C switch outlet. The D switch is provided to ensure a good grade of service on terminating calls without upsetting the modular construction of the A, B and C switches. All crosspoints have four reed inserts.

3.3.2 The supervisory and register area

Only three basic types of supervisory are needed although variations may be introduced to meet particular requirements. Each type contains a transmission and signalling bridge and appropriate logic circuits.

The outgoing supervisories provide facilities for routing dial pulse signals to a register, for onward transmission to other exchanges, for the receipt of called subscriber answer signals and service tones. They also generate metering signals in response to answering or, where appropriate, repeat metering from the called exchange.

The incoming and local supervisories are also arranged to receive dial pulse signals and to route these to a register. Provision is made for ringing the called subscriber's bell, the transmission of ringing tone to the caller and detecting the answer signal.

The registers associated with the supervisories are general purpose registers. Access is obtained via a register access switch (reed relay). The registers may be arranged to receive and store either dial pulse signals or multi-frequency coded push button signals.

3.3.3 The control area

The control apparatus is common to the exchange and is shown in block schematic form in Figure 3.1. It is duplicated to ensure continuous service in the event of faults. The two controls are independent and are used alternately with a periodic changeover to ensure that both controls are exercised and that faults will not go undetected for long periods.

When a customer or incoming junction originates a demand for access to a register, this demand is first routed to a calling number generator which identifies the calling line or junction and stores this number in a free register. The register then takes over and applies for access to call control to set up a connection from the calling line or junction to itself. The register marks all free supervisories of the first choice route and, simultaneously, sends the calling line's number to the called line marker. This decodes the number and marks the line concerned. The path selection apparatus now selects and establishes a path between the calling line and a free supervisory with access to the register already allotted.

This method of control greatly improves flexibility in use, since it takes advantage of the high speed of the system and permits originating calls to be released and set up again to a different point within the inter-digital pause and without the knowledge of the caller.

3.3.4 System operation

Each customer's line is connected via the MDF to a line circuit providing the exchange line termination. Junctions are terminated on supervisory equipments.

Each line circuit or incoming supervisory is given an identifying code known as the equipment number (EN) which indicates its location on the equipment racks.

To originate a call, the customer loops his line and causes his directory number (DN) to be inserted in a free register. If there are no free registers, the customer receives equipment engaged tone (EET) and is required to make another attempt later.

The allotted register applies to the call control for the selection of a free supervisory which, when found, causes the register to be connected via the register access switch, the chosen supervisory and the C, B and A switches to the calling line.

Initially, the allotted supervisory may be either an O/G junction supervisory or an O/E supervisory, depending on the facilities required and the distribution of traffic. For example, if the majority of the traffic is to the local tandem exchange, an O/G supervisory on this route will be the first choice. If there are no free supervisories on this route, a second O/G group or O/E supervisory is allotted instead.

When the register sets up the initial connection, it determines from the class of service signal received if any special facilities are required, for example if the call is from a public call box. When the facilities have been provided, dial tone is returned and dialling can proceed.

From the digits received, the register determines whether the call is:

1. An own exchange call.
2. A local call to or through the tandem exchange.
3. A long distance (trunk) call.

Assuming the tandem route is the first choice, in case 1, the register proceeds after the first digit or, in the case of a linked numbering scheme, a later digit to release the initial set up, to call in a free O/E supervisory and to reset the call, all during the inter-digital pause. The calling customer is not affected.

In case 2, as soon as sufficient digits have been received to indicate that the call is to the tandem exchange, the register is released and dialling continues over the already established path.

In case 3, the call will be released and reset to a supervisory in the trunk route.

Because all calls are connected from a supervisory to a customer's line, terminating calls are set up in a manner broadly similar to originating calls. Typically when a call arrives on an I/C supervisory, its terminal identifying number is generated by the calling number generator and placed in a free register which then proceeds to connect itself to the indicated supervisory. When the wanted number has been received by the register it applies to the decoder to set up an equivalent EN marking. Connection from the incoming supervisory then proceeds, as already described, except that an additional stage of switching (the D stage) is involved.

3.4 The TXE4 system

TXE4 is designed for local exchanges ranging from 2000 to 40000 lines and with calling rates from 0.02 to 0.35 erlangs per line. The system is readily extensible up to an overall limit of approximately 10000 erlangs of bothway traffic.

The system uses a reed relay switching network with electronic stored program control. The fundamental concept underlying the design is that it should consist of an assembly of a limited number of types of sub-system, hardware or software, arranged so that the exchange capacity in terms of connections, traffic or facilities can be extended by simple addition of similar subsystems. This arrangement illustrated in Figure 3.2 also ensures that exchange security is safeguarded in the event of failure of any subsystem without the need for interchange of information or external monitoring of the working of each subsystem.

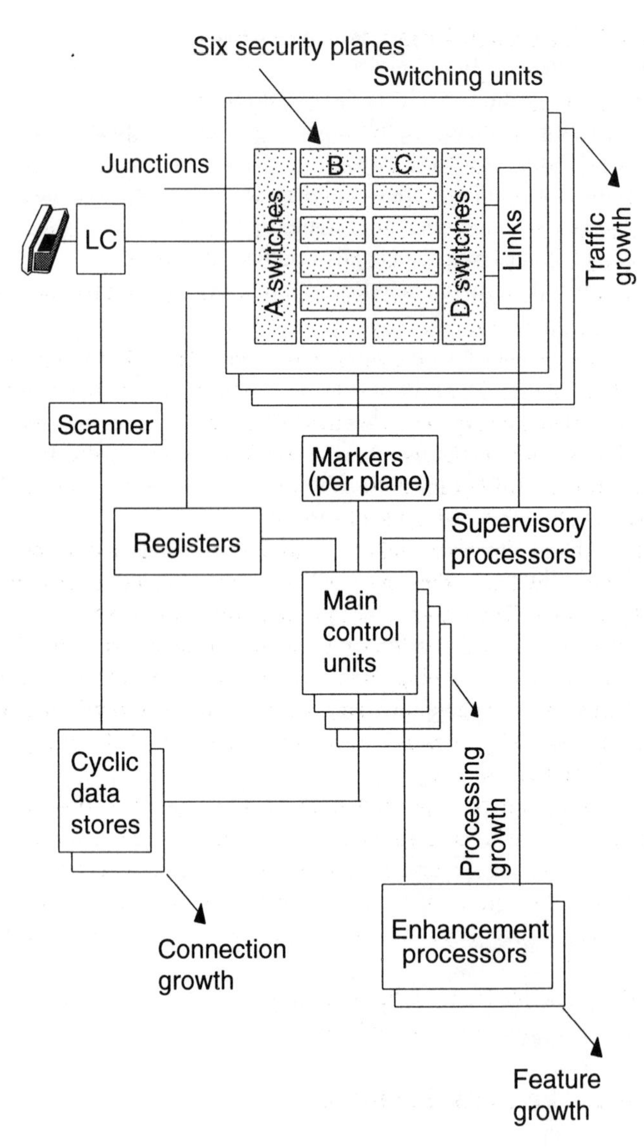

Figure 3.2 TXE4 system. (Compare with Figure 1.8)

3.4.1 The switching network

The switching network consists of a number of separate but identical units up to a maximum of 48. Each unit handles a fixed amount of bothway traffic and the number of units provided for any particular application is, therefore, determined by the total traffic (originating, terminating and tandem) to be handled. The network is arranged as a three stage link system folded upon itself via an additional switching stage, so that connection between any two terminations can be completed by the operation of seven crosspoints.

The BC stages of each switching unit are sectionalised to provide a number of identical sub-units (6 or 8) which can be considered as being in planes on top of one another. This concept of sectionalisation provides security in the switching network, since a fault affecting any one sub-unit will not mean the loss of service to any customers but only a slightly reduced grade of service.

Customers, junctions and other circuits are connected to the network via A switches, separate switches being provided for customers and junctions. The traffic concentration within the A switch can be adjusted to match the calling rate of the terminations to the fixed capacity of the switching unit. On the opposite side of the BC networks, interconnection is achieved by means of a fourth stage (D) of switching and groups of link circuits which provide local supervision where necessary.

The adoption of a general purpose switching network, as described, avoids the need for auxiliary switches for the connection of registers and other circuits. All such peripheral circuits are terminated on the general purpose network via junction A switches. Any peripheral may be associated with any termination by a path (or paths) through the network.

This principle leads to the concept of serial connection of switching paths known as serial trunking.

3.4.2 The control equipment

The control equipment supervises the setting up of all calls through the switching network.

3.4.2.1 *Cyclic data store and scanning*

The data stores provide common central storage for the exchange data relating to customers, junctions, code translations and routing information etc. The stores consist of read only shift registers which continuously circulate data and present this on common highways to all main control units.

Associated with the data store is scanning equipment, including additional storage for maintaining a record of the state-of-line information for customers. This is synchronised with the main data store so that all information relating to a particular termination is available simultaneously.

The basic store module provides for 5760 customers and a proportion of incoming and outgoing circuits. Up to seven data stores can be provided in an exchange giving a maximum size of over 40000 lines. For security, the complete shift register data files are duplicated and parity coding is used to enable single errors to be detected and corrected without any effect on exchange operation. A complete backup store is also provided on magnetic tape and this can be used to reload the data files.

3.4.2.2 *Main control units and registers*

The main control units are special purpose processors employing stored program control techniques. Up to 20 of these units may be provided dependent on the number of calls to be handled. The processors operate simultaneously but independently in a load sharing mode so that, in the event of a fault, the processor concerned is busied out and the traffic is distributed over the remaining processors at only a slightly reduced grade of service.

The overall operation of the main control processor and the manipulation of the information within it is controlled by a stored program held on read only memories. The use of non-volatile memory for program storage gives complete security against malicious or accidental corruption of the program. Changes are not required in service since the essential variable parameters are located in the data store with keyboard access.

Registers are processed in groups by the main control units, so that dialled digits from the calling termination, which are detected by the register, are stored in a common working store within the main processor. The register itself contains only the limited amount of equipment required to count the dial pulses under microprocessor control and to supervise the call during set up. Both 10 pps dial pulses and MF key signals can be accepted. The registers are connected to the periphery of the network via junction A switches, so that all registers are fully available to all customers and a standard setting up procedure can be used.

3.4.2.3 *Interrogator markers and supervisory processors*

Interrogators and markers are provided for each sub-unit of the switching network, although each marker can control up to five interrogators on the same plane. When path selection is required, the main control unit signals to all interrogators through their controlling markers and the free paths available are identified. The main control unit then selects the most appropriate route available and instructs the particular markers to operate the crosspoints within the selected sub-units and also to allocate the required type of link circuit for the connection.

The supervisory processor controls both the bridge link used for supervising local and terminating calls and also the outgoing junction circuit used for supervising outgoing calls. Each supervisory processor is triplicated for security and can control over 6000 connections at a time, serving 16 switching units. Data is obtained by monitoring the highways between the main control unit and the markers. The supervisory processor is program controlled using reprogrammable read only memories.

3.4.3 System operation

3.4.3.1 *Connection to a register*

When a call is originated either from a customer or an incoming junction, the calling condition is detected via the scanning by one of the main control units (MCU) which has been pre-allotted. At the same time, the equipment number, directory number and class of service relating to the calling line, available on the cyclic store highways, are stored in the MCU.

Under the control of its program, the MCU analyses the class of service of the calling line to determine if any special requirements apply, such as calls barred. If the call is to proceed, the MCU initiates a search over all registers available to it on the cyclic store highways. A free register is selected and its equipment number stored by the MCU in readiness for setting up the connection.

The MCU then seizes the common highways to the interrogator markers when these become free, thus busying them against seizure by other MCU. The equipment numbers of the calling line and register are then sent to all interrogators to identify free paths through the switching network. Interrogators with no free paths available release. Each of the others identifies the optimum path through its own sub-unit and signals this to the MCU which selects one of these for the actual connection. The appropriate markers are then instructed to set up a path. The register will signal to the MCU that the path has been completed and the MCU can then release. Should this path not be completed successfully, a second attempt is made automatically and the identity of the faulty path is printed out.

When sufficient information has been received by the register, the MCU is recalled and examines the dialled information to establish the destination of the call.

3.4.3.2 *Local call*

For a local call, the MCU stores the directory number of the called line. It then inspects the output of the cyclic store highways until by comparison it sees the same digits that it has stored.

At the same time, the state-of-line information indicates whether the wanted line is free or busy. If it is free its equipment number and class of service are stored by the MCU, which will then apply with the equipment numbers of both calling and called lines to the interrogator markers for final path selection. This is done in the same way as for the connection from calling line to register. Again, the successful completion of the path will be detected by the register, after which the original path between calling line and register is released.

Should the called line appear busy when it is scanned, the MCU immediately hunts for a free tone circuit and connects this to the calling line via the network to return busy tone.

3.4.3.3 *Outgoing call*

When the MCU examines the dialled digits received by the register and identifies that the call is outgoing from the exchange, it will, where appropriate, apply with the initial digits to the cyclic store for a code translation and will store any routing digits which may be returned together with the identity of the route and, where required, the number length.

The MCU then carries out a further search over the cyclic stores to identify a free junction within the required route and when this is found by comparison, its identify is stored within the MCU. Two paths are now set up in sequence in the same way as already described, one between the sending side of the register and the selected outgoing junction, and the other between the calling line and the outgoing junction, the latter path having a split function in the link circuit.

The MCU then instructs the register to send the appropriate routing digits over the junction, followed by the numerical digits identifying the wanted customer. When this has been done, the split is removed from the final conversation path and the connections between the calling customer and the register and the register and the outgoing junction are released.

3.4.3.4 *Supervision*

Once a connection has been set up, the MCU is released for the setting up of other calls and the established connection is monitored and supervised by the supervisory processing unit (SPU) in conjunction with either a bridge link circuit for locally terminating calls or an outgoing junction circuit for outgoing calls. The bridge link circuits apply the necessary ringing current and tones required.

3.4.3.5 *Feature evolution*

The TXE4 system provides service to some eight million lines, approximately a third of the U.K. network and it has been necessary over the years to enhance the features provided to match those provided by the newer digital systems. This has been achieved by the addition of specific functional processors interconnected by an Ethernet local area network and connected to the TXE4 MCU and SPU.

This arrangement provides call logging facilities including both bulk and itemised billing; fast call set up using Common Channel Signalling System No.7 to other TXE4 or digital exchanges; and centralised management of the operation and maintenance features of the exchange. It also permits the provision of supplementary subscriber services, both exchange and network based, and including the possibility of digital access to voice and data services. This enhancement process, which will make TXE4 exchanges almost indistinguishable to customers from fully digital systems, is still in progress and is likely to continue for the life of the system which is expected to continue well into the twenty-first century.

3.5 System 12 overview

3.5.1 Introduction

The System 12 functions are concentrated in different module types which are connected to a digital switching network. Each modular function is controlled by its own processor. Therefore the System 12 switching system is a modular system with fully distributed control, using the latest microprocessor technology and large memory capacities. The switching network control functions in System 12 are dispersed throughout the network. This results in an end-controlled network arrangement in which the several individual control elements (microprocessors) associated with different network ports are able to set up digital paths through the network simultaneously without a need for some central network controller.

The distributed control implementation in System 12 results in a number of system characteristics representing direct benefits to both network operators and users, e.g. robustness against total system failure, capability for smooth incremental growth both in traffic and control capacity, high degree of system uniformity leading to the need for only a very limited set of printed board types to build up an exchange. All these aspects make System 12 an outstanding product in the world of telecommunications.

Finally, good indications exist that for the next major evolutionary step in switching, the so called Broadband ISDN, fully distributed control systems based on end-controlled switching networks are the obvious solution. In this sense, System 12 is already prepared for this transition.

3.5.2 System 12 evolution

The studies, which lead to the System 12 configuration, were based on two main premises:

1. Digital systems should from the start be designed not just for voice traffic but also for an ISDN environment.
2. New semiconductor technology (LSI, VLSI) became available during the 1980s providing ample computing power and massive memory storage capability.

During this period also the end controlled digital switching network (DSN) was conceived and its basic LSI component (switch port) designed.

By the end of 1981 a first system was installed in the Belgian network and the next year 4 additional System 12 exchanges were installed in Germany.

Subsequently, System 12 product versions were quickly introduced in several countries in parallel, first in various Alcatel n.v. home markets, then in export. The System 12's novel attractive architecture aroused interest with many telephone companies and administrations throughout the world. By now over 60 million lines of System 12 are in service in 26 countries.

System 12's distributed architecture allows it to cover a very broad spectrum of exchange applications with respect to both size and network hierarchical level. Its expandable digital switching network and distributed control provide economical solutions for exchanges ranging from very small to very large. With respect to network levels, System 12 may be applied as local, tandem, national and international transit exchange. Also combined local transit exchanges and integrated or stand alone STPs are possible.

The list of individual exchange sites clearly shows the variety of applications: from small and large locals via numerous transits and combined exchanges up to several international transit exchanges including the highest possible level.

3.5.3 Developments and the future

The digital switching System 12 has the flexibility that is necessary to allow economic extension in terms of the number of lines and new services. It also offers high fault tolerance and rapid isolation and identification of faults.

Since the initial introduction of System 12 in public telephone networks, the range of its features and facilities has been significantly expanded:

1. To extend the application range to very small sizes as required for rural networks, a Remote Subscriber Unit (RSU) was developed.
2. For national and international transit exchanges, a fully featured Operator Position System called 12SO, was developed.
3. As required for several advanced features and in view of the quickly approaching ISDN in many networks, the ITU-T No.7 common channel signalling system was introduced in System 12.
4. A further enhancement was the development of a centralised operations and maintenance facility for System 12 networks, called the Network Service Centre (NSC). The NSC greatly benefits from System 12's ITU-T No.7 capability since the No.7 links provide the basic means of communication between the System 12 exchanges and the NSC.

Additional proof of System 12's flexibility is its application in various specialised configurations or entire specialised networks as used by governments and large institutions. Examples are exchanges which deal specifically with freephone service, signal transfer points for ITU-T No.7 Common Channel Signalling, wideband System 12 exchanges handling up to 30 × 64kbit/s connections and cellular radio switching applications like GSM, PCS and CTM.

Development of ISDN feature capability for System 12 was started some years ago and already System 12 ISDN field trials were held in a number of countries.

A feature which has recently been gaining importance in Europe is Centrex (see Chapter 5). A combined ISDN/Centrex package for local and/or wide area usage is available. Also the necessary attendant positions and queuing systems are provided.

3.6 System 12 architecture

The System 12 flexible architecture, which allows extensions and new services to be introduced as required, can be used for the entire range of exchange applications:

1. Rural areas: remote subscriber units and small exchange configurations: up to 1024 lines.
2. Urban and metropolitan areas: medium to large scale local exchanges: 256 to close to 200000 lines.
3. Toll exchanges: up to 80000 trunk terminations.
4. Transit applications: tandem, toll, international, and inter-continental exchanges.
5. Network Service Centres (NSCs).
6. Service Switching Point (SSP) for intelligent network applications

All exchanges throughout the range provide a full set of modern services and features to subscribers and network operators.System 12 independent exchanges are designed to cover both individual and multi-exchange network applications. They provide a full range of maintenance and administrative features including the possibility of remoting man machine communication terminals or connecting the exchange to a Network Service Centre (NSC). Figure 3.3 shows the modular structure of System 12.

A Q3 interface is provided and the necessary Q-adapter functions to connect the Operation System to the TMN network.

3.6.1 System 12 hardware configuration

A System 12 exchange comprises a Digital Switching Network (DSN) and a range of modules and Auxiliary Control Elements (ACEs) which perform all the exchange control and signal processing.Each module comprises a standard Terminal Control Element (TCE) and a terminal which is specific to the function of the module.

Figure 3.4 illustrates the Terminal interface configuration.

3.6.1.1 *Digital Switching Network (DSN)*

The DSN is a combined time and space switching network that provides time and space selection at each of this switching nodes.

The DSN comprises standard Digital Switching Elements (DSEs), which are used for all stages of switching.

3.6.1.2 *Control Elements (CE)*

All CEs have similar hardware comprising a microprocessor, with associated memory and a Terminal Interface (TI).

Most important hardware modules are described in the following sections.

3.6.1.3 *Analogue Subscriber Module (ASM)*

The ASM, which can be equipped as a paired module, provides the interface for up to 128 subscriber lines which are terminated by line circuits.

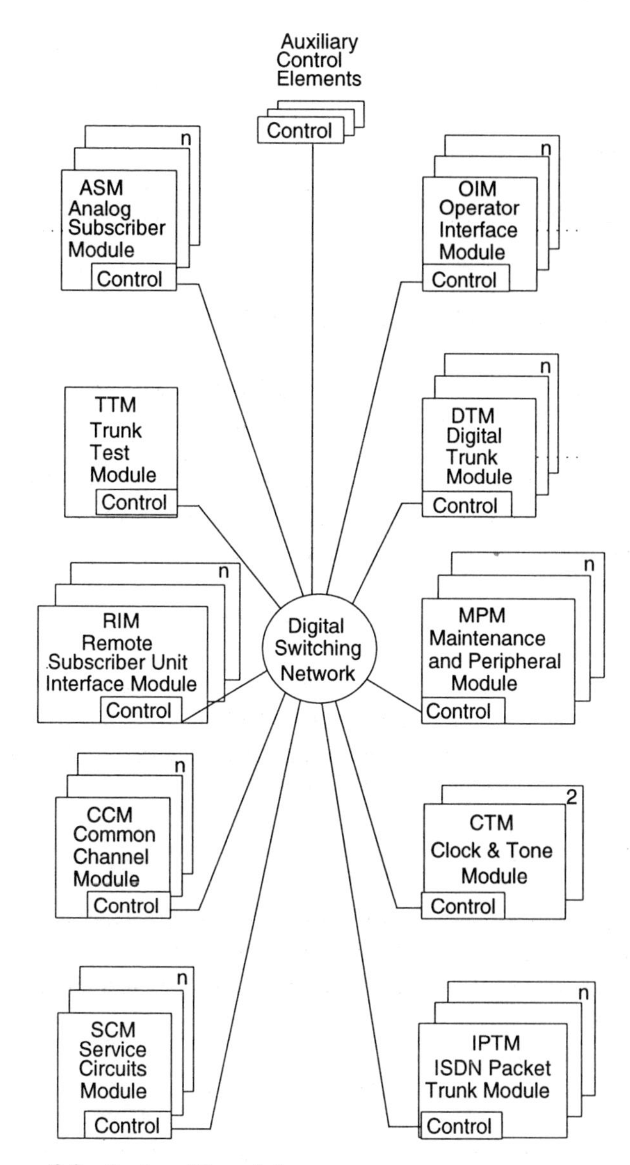

Figure 3.3 System 12 modules

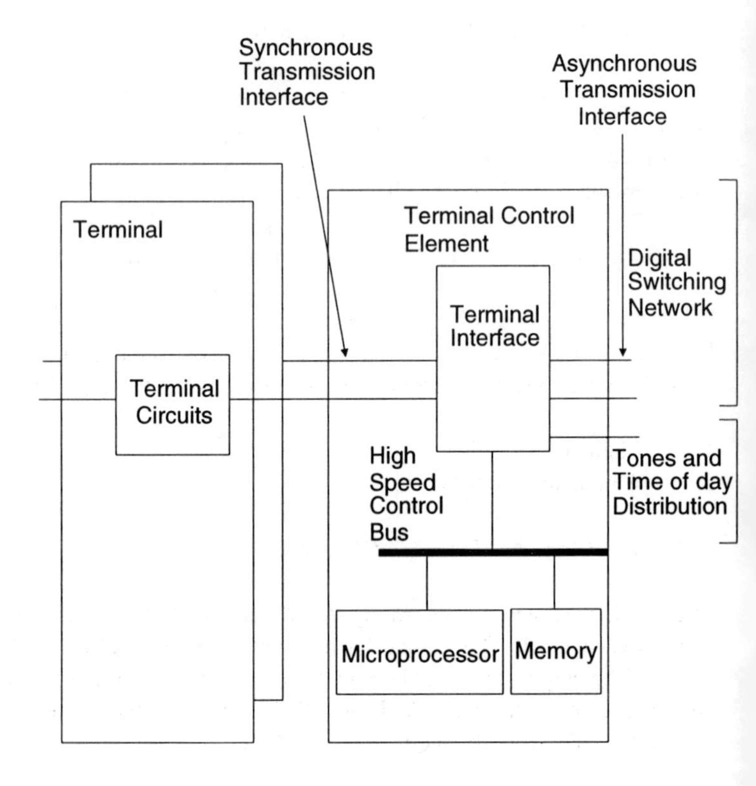

Figure 3.4 General structure of a Terminal Module

The module is controlled of two TCEs which are connected in a dual controlled pair configuration (also called the crossover mode).

3.6.1.4 *Digital Trunk Module (DTM)*

The Digital Trunk Module (DTM), performs the preprocessing and control for a single 30 or 31 channel PCM trunk.

The DTM converts the various PCM formats which can be used on a digital trunk to the standard format used in System 12 exchanges.

3.6.1.5 *ISDN Remote Interface Module (IRIM)*

The IRIM, which is equipped as a paired module, is similar to the DTM but is used to control Remote Subscriber Units (RSUs) and to interface them (singly or in multi-drop) to a host System 12 exchange. The lines can be analogue or ISDN lines.

3.6.1.6 *Maintenance and Peripherals Modules (MPM)*

An SPM, comprises a Support Module (SUM) and a Computer Peripherals Module (CPM), each of which has a TCE and a terminal. Each SPM is duplicated for reliability. There are two types of SPM:

1. Maintenance SPM.
2. Administration SPM.

3.6.1.7 *Clock and Tones Module (CTM)*

The CTM, generates a clock timing signal, a range of PCM encoded audio tones and the TOD signal for distribution throughout the exchange. The outputs of the CTM, which are duplicated for reliability, are distributed to the TCEs, ACEs and DSEs via a Clock and Tones distribution (CLTD) subsystem.

3.6.1.8 *Service Circuits Module (SCM)*

The SCM, provides tone handling facilities for subscribers with push-button telephone sets and also for exchanges that use multi-frequency (MF) senders and receivers for inter-exchanges signalling. Optionally, a conference call facility can be included.

3.6.1.9 *Common Channel Module (CCM)*

The Common Channel Module (CCM), performs signal processing for ITU-T Number 7 Common Channel Signalling (CCS) systems. The module can be used either for purely common channel signalling and for transfer of billing information, connection to IN, etc.

3.6.1.10 *ISDN Subscriber Module (ISM)*

The ISDN Subscriber Module (ISM), Provides up to 64 basic access (BA) connections for ISDN subscribers and controls the handling of related ISDN calls.

3.6.1.11 *ISDN Packet Trunk Module (IPTM)*

The IPTM can perform several functions always combined with the trunk module function.

Those functions are:

1. MTP functions for ITU-T N7 signalling.
2. PRA interface.
3. X.25 DCE functions for connection to the Network Service Centre.
4. Frame handler functions for D-channel multiplexing.
5. Packet Handler interface.

3.6.1.12 *Trunk Test Module (TTM)*

The TTM provides facilities for automatic, semi-automatic and manual testing of trunk devices.

3.6.2 System 12 general software configuration

The system 12 on-line software contains the following main areas:

1. Operating system and network handler, managing all exchange resources and control the SW in each CE.
2. Input/output software, interfaces between the application (user) software and the peripheral devices.
3. Man Machine (MMC) software provides a part of the interface between the MMC devices and the applications software.

4. Database software allows central management of data, common to a number of software modules.
5. The load and initialisation software provides initialisation and loading of software modules into all control elements during the start up or restart phase, initialisation and/or loading control of disc systems.
6. The maintenance software handles the processing of any exchange hardware and software faults to ensure the continuity of call handling operations.
7. The status and alarm software controls the status condition of terminal processors, records and displays alarm situations.
8. The Clock and Tone, calendar software contains the Clock and Tone device handler software, the associated diagnostic and routine tests, calendar scheduler handler and interactive control software.
9. The Line and Trunk Testing software gives test and maintenance facilities for subscriber and trunk lines.
10. The Call Handling and Facilities software interfaces with the other exchange to control the set up and connection of a call, through the exchange, to its destination.
11. The Signalling handling software provides an interface between the signalling systems and the call handling and facilities software.
12. The Telephonic Device and Signalling Adaption software provides the interface between the telephonic devices and the Signalling Handling software.
13. The Call Services software supports the Call Handling and Facilities software, manages the allocation of telephonic resources and it interfaces with the Database software (recalling data for prefix handling and routeing information).
14. The charging software performs call charging and accounting functions.
15. The administration software allows measurements, statistical and network management functions to be performed in an exchange.
16. Hardware and Software Extensions software enable extensions on hardware devices and updates of software functions.

3.7 System 12 O&M

Operations and maintenance (O&M) is the key to ensuring a high level of service quality. System 12 O&M facilities provide the following advantages for an administration:

1. Simple procedures for daily operations, such as line additions, trunk additions, trunk group assignments, changes, etc.
2. An effective charging system, for accurate and prompt billing, etc.
3. Easy exchange expansion or modification to satisfy increased traffic handling requirements or changing network conditions.
4. Measurement and Data collect procedures using simple Man Machine Command (MMC).
5. Efficient diagnostics and repair procedures, normally involving the replacement of a printed board assembly (PBA), which gives a low mean time to repair.
6. Minimum physical contact with the equipment, made possible by the use of the MMC facility. This facility is used to provide most of the O&M related functions, e.g. subscriber administration, corrective maintenance.
7. Easy modification of exchange data (routeing tables etc.) by the use of MMC commands. Modified tables etc. are automatically stored in back up mass memory (e.g. disc) to ensure that an updated copy of the data is loaded into the memory of a processor if a reload is done.

New systems were developed to ease O&M tasks, as follows.

3.7.1 The Network Service Centre

It can be directly integrated with a System 12 exchange or used in a standalone mode. It is constructed from standard System 12 modules, ITU-T No.7 signalling modules, and X.25 packet interface modules. Generally a network service centre is connected to a number of exchanges via ITU-T No.7 links.

3.7.2 Advanced Terminal for Operation and Maintenance (ATOM)

A user friendly man machine interface has been provided by means of PC programmes called ATOM.

This terminal provides functional screens representing parts of the system functions like e.g. system diagram, trunking diagram, floor plan etc. which allow by clicking of the mouse to instruct the system which function has to be performed.

Also for an easier overview of the exchange performance, post processing function are performed to display information like statistics, received error reports, alarms etc.

3.8 The AXE system concept

3.8.1 Introduction

The AXE system from Ericsson is widely deployed throughout the world. Today, AXE exchanges are in service in local, long distance, international and cellular mobile networks in some 80 countries.

AXE supports a complete range of telecommunication network applications. It is fully digital and designed to minimise overall network and handling costs. Stored Programme Control provides for an open ended supply of new features and services.

A unique modularity structure is the corner-stone of the AXE system concept.

3.8.2 System structure

The philosophy behind AXE is embodied in one word — modularity. Modularity means easy handling and ability to adapt to the changing world of telecommunications. Throughout the system, in both hardware and software, modularity is the guiding principle. This facilitates software and hardware design, system upgrades, system extensions and multi-functionality in network applications.

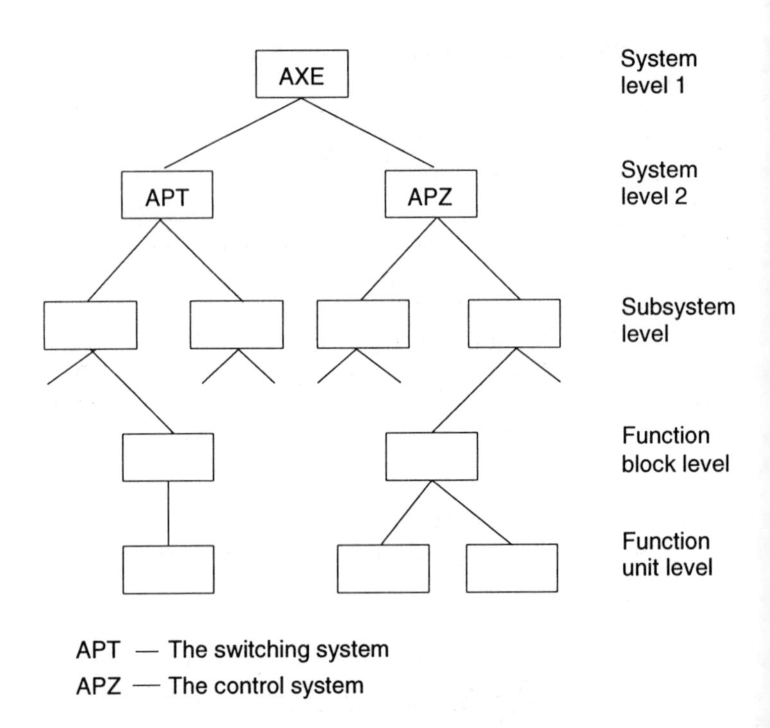

Figure 3.5 Functional levels in the AXE system

The AXE architecture is a five level hierarchy, each level made up of specific modules, as in Figure 3.5.

3.8.3 Functional modularity

Rapid developments in electronics technologies and of telecommunication networks have imposed specific requirements on switching systems. Functional modularity is a prerequisite if the system is to be introduced easily and effectively in a variety of different and changing environments. AXE is defined in terms of functions; what they do rather than how they are to be implemented. Functions will not change in the future; implementation techniques may.

The entire AXE system is a set of specified functions, implemented in modules, or function blocks. These function blocks are grouped in subsystems. Each function block and subsystem is considered as a 'black box' at its specific level in the system hierarchy, and has a defined interface to other function blocks and subsystems. Since each function block is defined by its interface with other functions, it is of minor significance how the function is implemented, in hardware and/or software.

The heart of the AXE system structure is the function block with its function units. The function units are located within each function block. Here, the distinction between hardware and software becomes evident. A typical function block consists of a hardware unit, a regional software unit and a central software unit (see Figure 3.6).

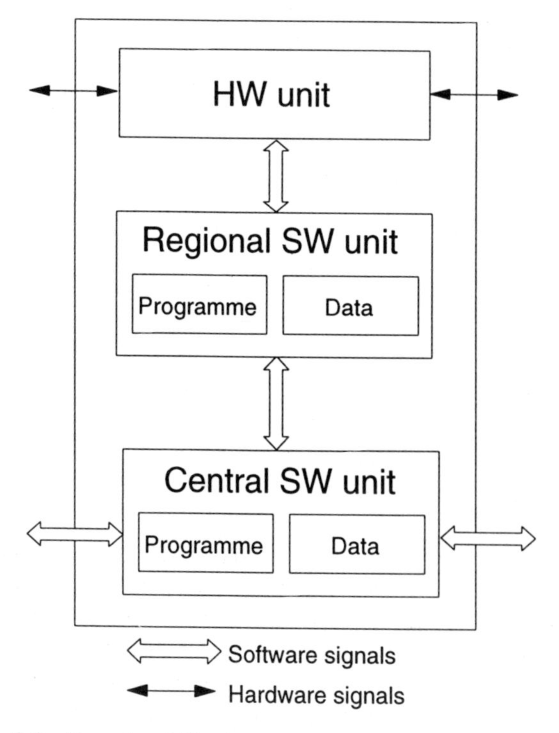

Figure 3.6 Functional block structure

The interwork between function blocks and function units is carried out by means of software signals. Software signals are a set of data which precisely define the interface between the 'black boxes'. The use of software signals permits design modifications with minimum overall system impact, that is, without affecting other function blocks. A typical AXE telephone exchange is built up of 400 – 500 function blocks.

3.8.4 Software modularity

Each software unit is programmed independently of all the others. Interaction between software units is by means of software signals and each software unit can only work with its own data. Consequently, a programme fault in one software unit can not mutilate data belonging to other software units. In other words AXE provides inherent software security.

Hardware units normally exist in numbers of identical devices with the exchange. In a similar manner, regional software units are repeated in regional processors controlling a certain type of hardware unit. The central software units are stored in the central processor.

Figure 3.7 illustrates the internal structure of a central software unit. The software unit consists of a programme part, which contains

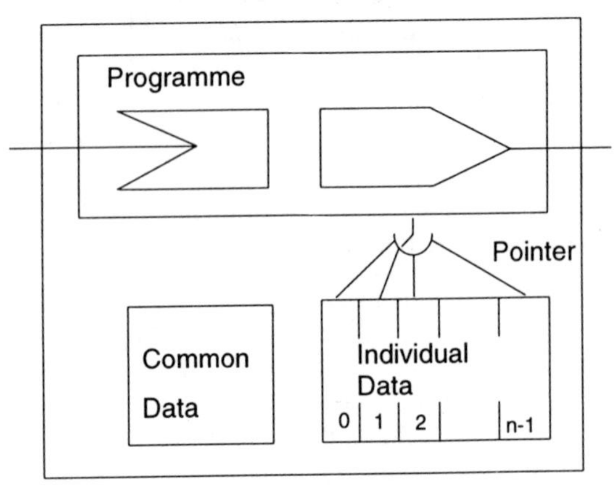

Figure 3.7 Central software unit structure

the logic and the ability to receive and transmit software signals, and a data part, which contains the stored information of the unit. This information may be common or can be individual for the controlled hardware units. In the latter case, the current data record is addressed by a pointer.

To obtain software security and ease of handling, the control system supports this functional structure with:

1. Special microprogrammed machine instructions used to transmit software signals between central software units.
2. Protection against data access to data of other software units. In addition, all absolute address calculations are performed automatically by microprogrammes.
3. Relocatable central software units; as the programme and data parts of a software unit can be stored in any free area of the programme and data store, each software unit can be handled separately, when faulty software units are replaced, or when new functions are introduced to the system.

Each function block in an exchange is assigned a unique function block number. Only when the function block number register (one of the programme control registers) contains the function block number for a certain block, access can be obtained to the programme and data.

Access to data which is only permitted within the function block uses a base address method, in which every separate piece of data is given its own base address. The base address is found in the reference store. This base address contains all the information about the location of the data in the data store and how addressing is carried out. In this way address calculations are made in hardware and not in software, which simplifies programming considerably. By using this base address method, all data in the data store is completely relocatable, that is, it can be placed in or moved to any idle area in the data store.

The programme for a function block forms a continuous area in the programme store. The start address to this area is also found by means of the reference store. The machine instruction list does not contain any operations which refer to an absolute address in the stores. Thus the programme part of a function block is also relocatable.

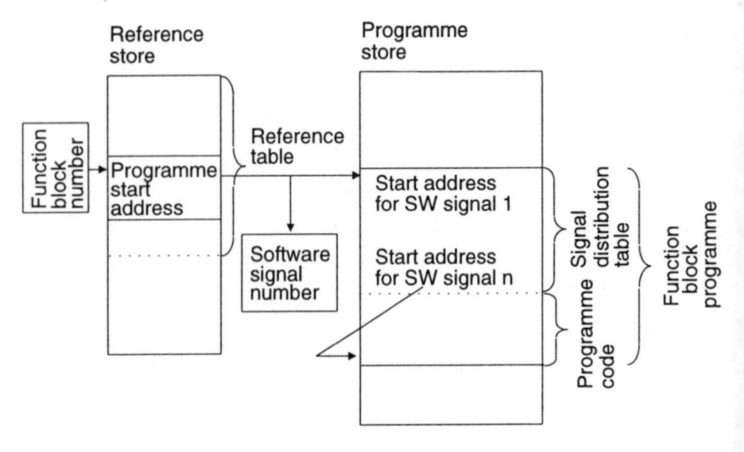

Figure 3.8 Addressing principle

Each software signal to a function block is defined by a unique symbolic number. In the software signal operation the block number for the receiving function block and the signal number in question must be given in order to access a specific programme sequence, (See Figure 3.8). First the function block number register is loaded with the block number. The function block start address is then extracted from the reference store and finally the symbolic signal number is used to obtain the entry address to the task sequence from a signal distribution table located in the beginning of the programme. In this way, programme changes made in a function block do not affect other function blocks as long as the software signal interfaces are maintained.

The major part, i.e. all central software units, of the programmes for AXE are written in a high level language, PLEX (Programming Language for Telephone Exchanges) that takes the modularity attributes, function blocks and software signals into account.

3.8.5 Hardware modularity

AXE hardware is composed of magazines, mechanical frames holding printed board assemblies, which are housed in cabinets. Hardware

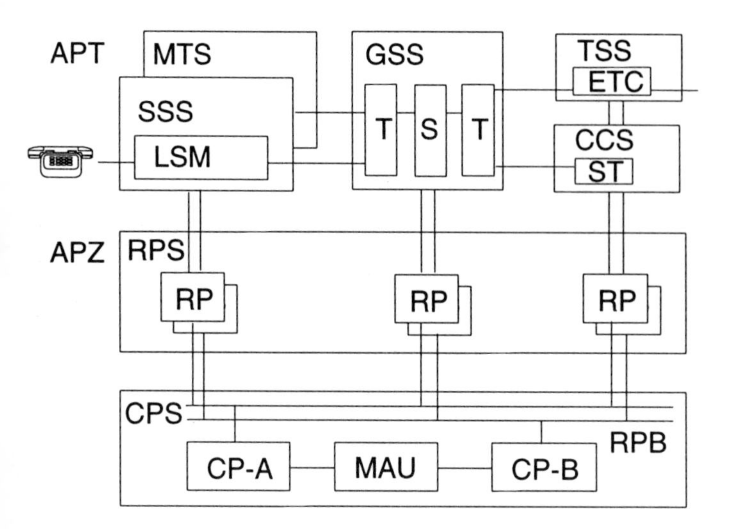

APT Switching System
MTS Mobile Telephony Subsystem
SSS Subscriber Switching Subsystem
GSS Group Switch Subsystem
TSS Trunk and Signalling Subsystem
CCS Common Channel Signalling Sub.
LSM Line Switch Module
ETC Exchange Terminal Circuit

APZ Control System
RPS Regional Processor Subsystem
CPS Central Processor Subsystem
RP Regional Processor
CP Central Processor
MAU Maintenance Unit
ST Signalling Terminal

Figure 3.9 AXE hardware structure

units can be added, modified or deleted while the AXE system is operational. All connections between frames are made by plug in cables. The AXE hardware structure is shown in Figure 3.9.

From the smallest component on the printed board assembly up to the cabinets containing full equipment, a building block philosophy is applied.

3.8.6 Technological modularity

Technological modularity, in turn, permits the adoption of improved technical components within the framework of the system.

New technology, e.g. new components, may be introduced, making possible a continuous development of the system. AXE therefore always represents state of the art technology.

3.9 AXE implementation

One system for all applications is achieved through specified functions, grouped together to form subsystems. These may be combined to suit different applications. An AXE exchange is a combination of switching subsystems and control subsystems.

3.9.1 Switching system APT

The switching system APT performs traffic handling functions and also operation and maintenance functions related to traffic handling.

The division of APT into subsystems is determined by conditions and requirements that arise from traffic handling and operation and maintenance functions. These subsystems are implemented both in hardware and software.

In the following sections the basic APT subsystems are described.

3.9.1.1 *APT subsystems*

The subscriber Switching Subsystem, SSS, supervises the state of connected subscriber lines, sets up and releases connections in the subscriber switching network and sends and receives signals to and from subscribers. SSS is a mandatory subsystem in local exchange applications.

The SSS hardware, the digital subscriber switching stage, consists of a number of Line Switch Modules (LSM) each of which contains line circuits for 128 analogue subscriber lines or alternatively 64 digital subscriber lines (2B+D). Special LSMs also exists for alternative accesses to the subscriber network, such as ISDN 30B+D. A hardware block diagram of the digital subscriber switching stage for analogue accesses is shown in Figure 3.10.

LICs, KRD and JTC/ETCs are connected to a time switch (TS). The time switch is non-blocking and has full availability. Thus any

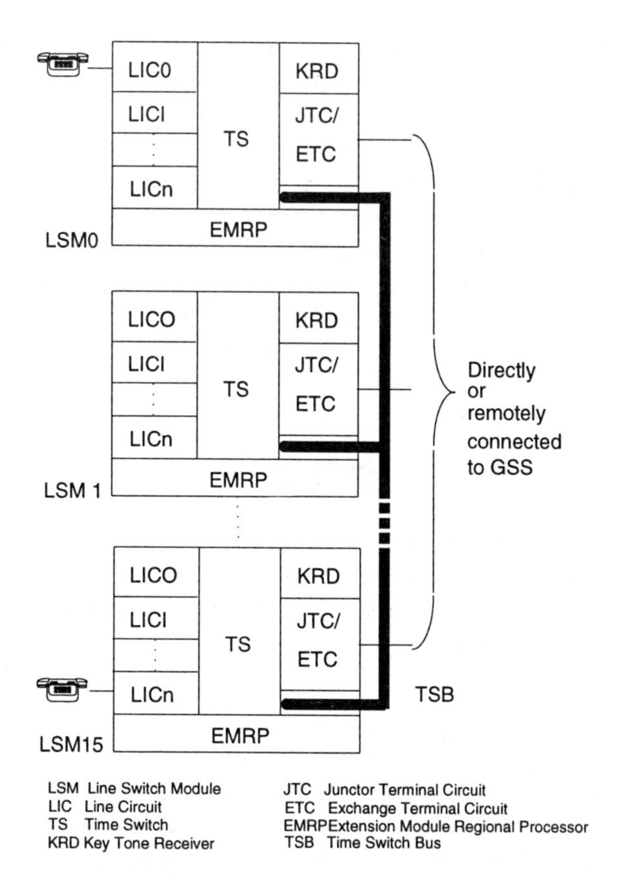

Figure 3.10 The digital subscriber switching stage

channel in the JTC/ETC within an LSM can be reached from any LIC within the same LSM. Via a time switch bus (TSB) any channel in other JTC/ETCs can also be reached. Consequently, the access switch is completely insensitive to unequally distributed loading (mixture of high and low traffic subscribers).

When the subscriber switching stage is remotely located, built in emergency functions are activated in case of a break in the communication towards the host exchange. These functions allow the subscriber to make internal calls with reduced facility level. It is also possible to connect one of the subscriber lines to an emergency centre in order to maintain '999' service.

The Group Switch Subsystem, GSS, is used to set up speech paths between connected devices.

The Digital Group Switch has a T-S-T (time-space-time) structure. It is built up of a number of time switching modules and a number of space switching modules. Time switching is performed by buffer memories and space switching by electronic crosspoint matrices. The switching part of the group switch is duplicated in two planes and all calls are established in both planes. The active plane is determined on a per call basis.

The timing in the digital group switch is performed by a triplicated clock module. Synchronisation is implemented by means of a software algorithm that can accommodate alternative network synchronisation methods.

The Trunk and Signalling Subsystem, TSS, supervises the state of trunks to and from other exchanges. For analogue and analogue to digital conversion trunks, the signalling is extracted and inserted in the trunk before entering the switch. Multi frequency signalling is handled via the group switch using code senders and code receivers.

The Common Channel Signalling Subsystem, CCS, implements the Message Transfer Part (MTP) of the ITU-T signalling system No.7.

Signalling system No.7, as specified by ITU-T, is built up of two parts, namely User Parts (UPs) and a common Message Transfer Part (MTP). The task of the MTP is to transfer information reliably between UPs in different switches. The user parts belong to the Trunk and Signalling Subsystem, TSS.

The Traffic Control Subsystem, TCS, controls and supervises the set up of all calls. TCS stores and analyses digits received directly from subscribers or via trunks and based on stored information (analysis data) decides how the call is to be handled.

The Charging Subsystem, CHS, performs the task of charging calls or at least to collect raw data needed for charging purposes. The charging system can be used for either pulse metering, itemised billing or both.

The Operation and Maintenance Subsystem, OMS, is equipped with a wide range of functions for supervision, fault locating, statistics and administration. AXE is designed to work without any external support system but necessary interfaces to interwork with centralised operation and maintenance systems are also provided.

3.9.2 The control system APZ

The control system architecture, with a central processor and a varying number of regional processors, is optimised for real time data processing. The real time environment imposes high demands on the control system: calls appear in an irregular manner; a variety of call models must be supported; very short response times are required; and overload situations must be handled.

All the specific requirements and demands of real time data processing are efficiently handled by the AXE distributed control solution; an expandable front end line up of small but fast processors, controlled and co-ordinated by a central processor.

Frequent call processing functions such as scanning operations and signal processing present stringent real time requirements. In AXE these functions have been assigned to special purpose regional processors. The central processor is free to handle the higher level processing functions. The two processor levels are shown in Figure 3.9.

3.9.2.1 *APZ subsystems*

The Central Processor Subsystem, CPS, consists of a duplicated central processor (CP). The APT central software units are stored and executed in the central processor.

The two processor sides work in a synchronous parallel mode. Permanent or temporary (intermittent) hardware faults can affect only one side at a time, data mutilation occurs in that side only. The intact side automatically continues traffic handling with correct data while the faulty side is taken out of traffic. This process is sufficiently fast to make corrective software actions unnecessary. Thus, in the majority of cases hardware recovery will not affect switch operation.

The synchronous duplication model in fact guarantees that central processor hardware faults will not cause software failures.

In addition, comparison between the two processor sides is an effective diagnostic tool for locating a faulty hardware sub-unit. The ability to separate the system without affecting its operation is also a valuable aid during major extensions or substantial software modifications.

Error correcting techniques in the memories mean that most memory faults are corrected without affecting the application software.

AXE can be equipped with either of two alternative central processors that are fully compatible on target code level, towards RPs and I/O peripherals and may be changed or modified, even in service. One is aimed for low cost, small physical volume and high reliability whereas the other is aimed at high capacity requirement applications. They have a similar logical and physical structure. Each processor stores software in three logical stores. Programme Store (PS), Data Store (DS) and Reference Store (RS).

The operating system is structured to be largely hardware independent. It has functions for job monitoring, loading, function change, automatic re-loading, storage allocation, size alteration of data files, programme correction, programme test, processor statistics and handling of regional processors.

The Regional Processor Subsystem, RPS, consists of a dimensionable number of regional processors (RPs). The APT regional software units are stored and executed by RPS. The interwork between the central processor and the regional processors take place via a duplicated regional processor bus, RPB.

The Maintenance Subsystem, MAS, supervises the proper operation of APZ and takes appropriate action should faults occur. The MAS functions can be divided as follows:

1. Supervision for detection of occurred faults.
2. Tasks after fault detection to limit the effect of a fault and to localise it.
3. Repairs, including methods and aids to execute repairs with minimum disturbance.
4. Support for CP hardware extension.

Finally there are four subsystems designated for input/output functions. To relieve the central processor, input/output functions are carried out by a support processing system connected to the central processor.

The Support Processor Subsystem, SPS, contains the above mentioned support processing system. It is built up of one or several general purpose type processors that can be configurated in a modular way, from small to large configurations, depending on capacity and reliability requirements.

The Data Communication Subsystem, DCS, consists of support processor software for data communication. It provides various kinds of data communication facilities supporting remote operation and maintenance, transfer of charging and accounting data to a remote or local facility etc.

The File Management Subsystem, FMS, comprises of support processor software for advanced file management. FMS supports various file structures such as direct, keyed or sequential access. Files may also be transferred to a data link via DCS. File transfer normally takes place via hard disc, which function as a backup for the data link.

The Man Machine Communication Subsystem, MMS, consists of central and support processor software, and hardware devices for alphanumeric input/output. Man machine communication in AXE is performed via alphanumeric terminals and alarm displays. The man machine language is consistent with the recommendations of ITU-T. Terminals can be placed locally or remotely and remote subscriber stages can also be fitted with transportable terminals when needed. Unauthorised access is prohibited by a number of security systems, e.g. authority checks, passwords and identity cards.

3.9.3 Packaging structure

The AXE packaging structure has been developed to meet the requirements of a modern switching system. From both mechanical and electrical viewpoint, the packaging offers a very high degree of flexibility.

The basic unit is the magazine. Each magazine corresponds to a function of the AXE switching system. At the factory the magazines are fitted with their printed boards assemblies. Then they are tested as functional units, either as single magazines or as complete cabinets. Accordingly, a complete switch may be pretested in the factory thus reducing on-site activities to a minimum.

The packaging system contributes to easy handling all the way: during design, manufacture, installation and operation/maintenance.

3.10 AXE applications and features

The concept of AXE is to provide one system for all network applications through a system wide concept of open ended modularity.

Telecom operators request services to be tailored to their individual needs and require some degree of control over the services. The opportunities for creating new sources of revenue for operators seems tremendous. Successful exploitation of these new opportunities will depend on getting new services to market very quickly.

3.10.1 The residential network

In the residential network AXE can be used for:

1. Local exchanges for metropolitan areas, towns and rural areas.
2. Remote subscriber switches.
3. Local (tandem) and national transit exchanges.

AXE local exchanges comprise a variety of subscriber services. Typical services are abbreviated dialling and hot line, multi party conference, call transfer, call barring etc.

3.10.2 The business communication network

The different activities within large organisations are normally distributed between several widespread locations nationally and internationally. In fact, structures with increasing geographic distribution can be envisaged, such as offices located remotely or even in the home. By applying a combination of soft (additional software to already installed local exchanges) and hard (new switches) network overlay principles, all locations within a company can be interconnected via a Virtual Private Network (VPN).

AXE provides a complete business communication package with Centrex type services which combines all the functions required by the demanding business subscriber: implementation in any network application; both mobile and wired access; provisioning of both voice and data services; service to both large and small companies; coverage of single site as well as multi site companies.

AXE comprises a full set of ISDN services in accordance with the ETSI Memorandum of Understanding. ISDN accesses can also be incorporated within the business groups.

The first generation of AXE broadband switches will be based on pre-B-ISDN offerings, such as Local Area Network (LAN) interconnect and virtual leased line services for High Speed Data Communication (HSDC).

3.10.3 International gateways

International telephony involves many special demands: high traffic load day and night, demands for secure charging and accounting, demands for sophisticated maintenance functions, advanced operator services, and a full range of national and international signalling systems. AXE offers a complete international exchange system, including large trunk switching capacity, a powerful central processor, all international signalling systems, advanced charging and accounting functions, and an operator system.

3.10.4 The operator system

The operator system contains functions for operator handling of manually assisted calls. It can provide services for national and international switching centres and also for directory inquiry service centres, message service centres, alarm collection centres, etc. An operator position can be used for all types of traffic, thus allowing for an integration of operator assisted services.

The operator system includes automatic demand and delay ticket handling; automatic charging and accounting, same as in the overall concept; automatic call distribution functions, based on queues and priority levels, resulting in a maximum utilisation of the system in all situations.

The operator terminal system is based on standard personal computers and the terminals can be clustered by means of a Local Area Network (LAN). The operator terminals can be freely located in large or small office areas or even in the home environment. Through the LAN, operators may access a wide range of data bases.

3.10.5 Mobile networks

AXE can be equipped as a mobile switch for both European and North American standards for both analogue and digital mobile telephony. All the major analogue standards, NMT 450, NMT 900, TACS and AMPS, are supported by AXE as well as the digital standards for the pan-European GSM and the North American ADC networks.

In a mobile network the AXE switch can be used in the following applications:

1. As Base Station Controller (BSC).
2. As Mobile Switching Centre (MSC).
3. As Visitor Location Register (VLR).
4. As a transit switch interconnecting the MSCs and to provide access to the fixed network.
5. As a Service Control Point (SCP) for IN (Intelligent Network) functionality (the service control functionality can also be integrated on the MSC or transit layer).
6. As an operator exchange for manually assisted calls.

The mobile applications covered by AXE also encompasses functionality for Personal Communication Networks (PCN).

3.10.6 Intelligent Network (IN)

The AXE/IN architecture consists of six functional levels:

1. Basic switching.
2. Service switching.
3. Service control.
4. Network management.
5. Service management.
6. Service creation.

The first three layers are implemented in AXE. The last three in a specially tailored Service Management System (SMAS).

The central processor in AXE is by nature a real time, high availability computing system, optimised for use within telecommunications networks. It fits with the intelligent network application as it meets the highest transaction rate and short response time requirements.

As the AXE can be configured both as Service Switching Points (SSPs) and as Service Control Points (SCPs), the possibility to make combined Service Switching and Control Points (SSCPs) also exists. This approach allows for the service execution to migrate from a central service control point (SCP) into the local network (SSCP) once a service has become established and widespread.

The AXE/IN implementation has a built in software platform, the service script interpreter, which is generic, or service independent, across a range of services. This allows the telecom operator to design and develop new service software in the form of service scripts. The Service Management System provides the tool for service creation, and it also allows for service management and end user control of service data.

3.10.7 New switching techniques

In the transport network, the use of a wider range of switching techniques will be implemented in order to permit a more flexible allocation of bandwidth. Traditional circuit switching will be supplemented by Asynchronous Transfer Mode (ATM) techniques for services such as LAN (Local Area Network) interconnect and other high speed data communication services, finally forming the new infrastructure of the Broadband ISDN (B-ISDN).

Table 3.1 Changes in telecommunication requirements

From	*To*
Node defined switching	Network wide services
Copper bandwidth	Fibre bandwidth
Node intelligence	Intelligence diffused through the entire network
In band signalling	Common channel signalling
Standard services	Customised services
Wire access	Wire and radio access
Local/regional standardisation	Global standardisation
Monopolistic regulation	Competitive markets
Circuit switching	A mix of circuit and packet switching
Narrowband services	A mix of narrowband, wide-band and broadband services
Calling defined locations	Calling personal numbers
Plesiochronous transmission	Synchronous transmission

3.10.8 AXE development

AXE has been continuously further developed since the first exchange was brought into service in 1977, but the ever changing telecommunication arena constantly provides new demands. Table 3.1 gives some examples of already ongoing changes, including the possibility of digital access to voice and data services.

3.11 DMS system

The DMS (Digital Multiplex System) family of digital switches manufactured by Northern Telecom has variants designed to meet the requirements of all levels in the network hierarchy (Figure 1.2), e.g. DMS-100 for local exchanges of 1000 to 100000 lines, DMS-200 for tandem exchanges of up to 60000 trunks or junctions and DMS-300 for international gateways up to 27000 trunks.

Through the use of highly modular architecture, for hardware and software, Centrex, combined local/tandem, mobile telephone, military and PABX (DMS Meridian) versions are also available. This modular architecture also enables the Intelligent Network applications to be integrated with the DMS-100, DMS-200 or DMS-300 applications as appropriate.

The modular architecture permits customisation to meet local needs and DMS is in service in many countries world-wide including the USA, Canada, Europe and Japan. In the U.K., DMS exchanges form the basis of the Mercury network and provide advanced services in the BT network.

Members of the DMS family are characterised by their ability to meet new application and reliability requirements, and to incorporate new technology in their design. New technology in turn provides increased capability and reliability at reduced cost.

The DMS family can evolve because modular design techniques are used in the development of both its software and its hardware. Modularity may be thought of as the implementation of a complex system through a set of functional units or modules connected by well defined interfaces. As a result of proper module and interface design, the various units can be connected, disconnected, modified, or improved, without affecting either the operation of other modules in the system or the system as a whole.

There are two generations of DMS. The second generation is known as DMS SuperNode and provides increased processing and call handling capacity at reduced size and with improved reliability. This is the current product and, except where otherwise stated, the following description refers to DMS SuperNode. The first generation

(NT-40) is, however, in widespread use so its principal features are also mentioned.

An overview of the DMS system architecture is shown in Figure 3.11. In the first generation system the speech links are 4 wire copper

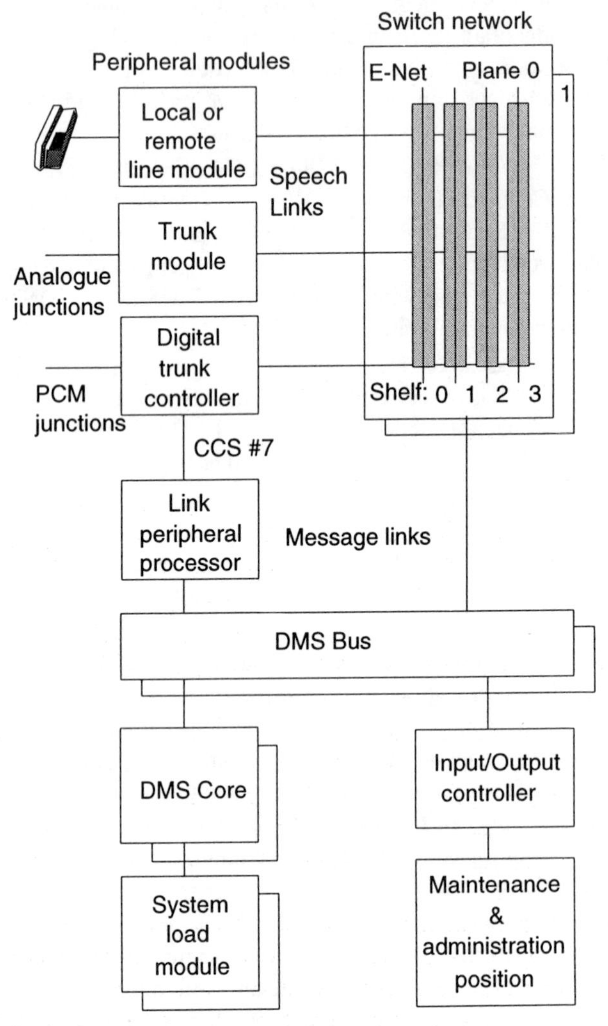

Figure 3.11 DMS (SuperNode) system

at 2.56Mbit/s in DS30 format comprising 30 digital speech channels plus parity and supervision. Second generation systems employ a mixture of DS30 and DS512 optical fibre links (equivalent to sixteen DS30 links each). Similarly, the message links are either DS30 or DS512 but with all channels used for message data.

3.12 The DMS switching network

The switching network provides two way digital speech and messaging connections between the Peripheral Modules (PM), interfacing both originating and terminating lines and junctions. Security of each network connection is achieved by providing two separate speech paths (Planes 0 and 1) for every call.

The first generation employs a Junctored Network (J-net) comprising a series of interlinked Network Modules (NM). Each NM provides one stage of time switching to route each call from one PM via a junctor link to another NM which routes the call to the required terminating PM. There can be up to 32 NM, each interfacing 64 DS30 links giving a network capacity of 64K channels.

The second generation network which is now being used for all new work is known as the Enhanced Network (E-net) and comprises a single stage junctor-less non-blocking switching matrix of up to 128K channels, as shown in Figure 3.12. The crosspoint cards comprising the matrix each contain four 16K (in) by 4K (out) time switches combined to form a 16K by 16K matrix. The entire E-net is housed in two cabinets, one for each plane.

Each shelf of crosspoint cards is controlled by its own processor communicating with the DMS-Core over a DS512 message link.

Referring to Figure 3.12, it can be seen that four incoming DS512 pcm signals, or an equivalent number of DS30 streams, are multiplexed into a single 2K (2048 channel) stream by each speech link interface card. The 2K pcm stream from each interface card is passed to the crosspoint card positioned directly in front of it and also distributed by a vertical bus (V bus) to the corresponding crosspoint cards mounted on the other three shelves. Each crosspoint card thus receives four 2K pcm streams, one from its own interface and three more from the interface cards above and below on the other three

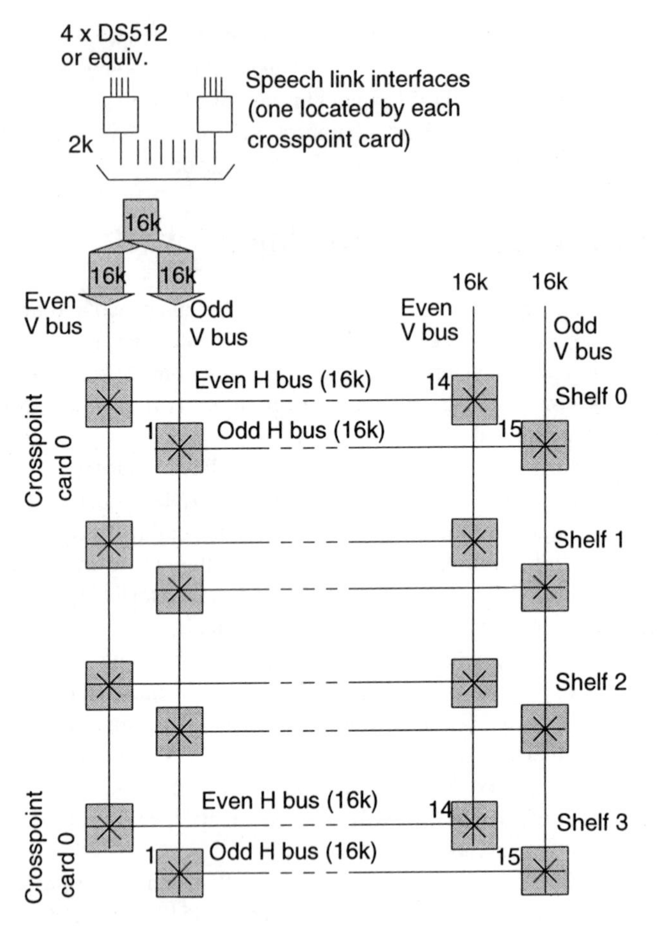

Figure 3.12 DMS switch network (E Net)

shelves, a total of 8K channels. In addition, the vertical buses are mated in odd/even pairs so that each card also receives a further 8K channels from the adjacent mated bus. This gives a total of 16K channels multiplied over 8 cards comprising an odd/even pair on all four shelves.

With up to 16 (8 pairs) crosspoint cards per shelf, this gives the total of 128K channels in eight streams of 16K.

Each incoming call will appear on eight of the crosspoint cards, serving the 16K stream within which it originates. One of these cards switches the call to one of the pair of horizontal buses (H-bus) available on its shelf. The call is then switched by the appropriate card on the same shelf and the same H-bus serving the 16K stream containing the destination port. Since any incoming call appears on both the odd and even card on each shelf via the mated V-bus arrays, it may be switched to any other crosspoint card via the appropriate H-bus. The second crosspoint card simply passes the incoming 2K stream from the H-bus to the speech link interface for demultiplexing into four separate DS512 speech link ports. A call between two channels in the same 16K stream is switched twice on the same crosspoint card.

In terms of the definitions of time and space switching, given in Chapter 1, this network can only be described as a time switch. There is no separation of time and space switching functions and the whole network can be regarded as a space switch having time switches as its crosspoints. Each call is connected via two time switches but the network itself is not divided into separate stages.

It will be seen that this switching network corresponds to the Group Selection stage of Figure 1.6. The line concentration is provided in the relevant Peripheral Module (PM).

3.13 DMS Peripheral Modules (PM)

A range of different PM is provided to interface the Switching Network to all external facilities. Four typical ones are shown in Figure 3.11 and these are described in the following sections.

3.13.1 Line Concentrating Modules (LCM)

There is a variety of these PMs which may be housed either at the Central Office (host site) or at a location closer to the customer premises being served (remote site). They support analogue lines and provide low level functions like line scanning and ringing.

A maximum of 640 lines can be connected to each LCM. Each LCM consists of two shelves, known as Line Concentrating Arrays (LCA). There can be up to 320 lines connected to each LCA.

A group of LCM connects to a Line Group Controller (LGC), a Line Trunk Controller (LTC), or a Remote Cluster Controller (RCC), which in turn connects to the Switching Network. There are a maximum of three DS30 links per LCA and a maximum of six DS30 links per LCM.

For reliability, each Line Concentrating Array is capable of taking over the lines of the mate LCA. The LCAs are connected by a serial data link that allows one LCA to checkpoint its data with the mate LCA.

The data for each call in progress is sent to the mate LCA over this link. If a fault occurs in one LCA, the mate LCA can take over the calls in progress.

Between the two LCA, there can be one or more speech links (DS30). If all channels on an LCA are busy, but the mate LCA has free channels, a call originating on the busy LCA can be routed over one of the inter-LCA speech links to a free channel on the mate LCA. This capability provides access for all lines to all six DS30 links for traffic engineering purposes.

The LGC performs high level functions, such as call coordination and the provision of the different tones required. It is equipped with duplicated processors operating in hot worker/standby mode.

As Figure 3.13 illustrates, there is usually some concentration of lines, depending on the engineering of the particular exchange. The concentration can occur in two places: in the LCM or in the LGC.

Each LGC has a maximum of ten LCM and, therefore, a maximum of 6400 (10 × 640) lines. The LGC has a maximum of 20 ports available for DS30 links from the LCM and, therefore, also has a maximum of 600 (20 × 30) speech links from the LCM.

There are up to 480 (16 × 30) speech links between the Switching Network and the LGC.

Line concentration can be lessened either by reducing the number of Line Concentrating Modules or the number of lines per LCM.

Remote line modules can provide a limited local switching service should the remote site become isolated from the host.

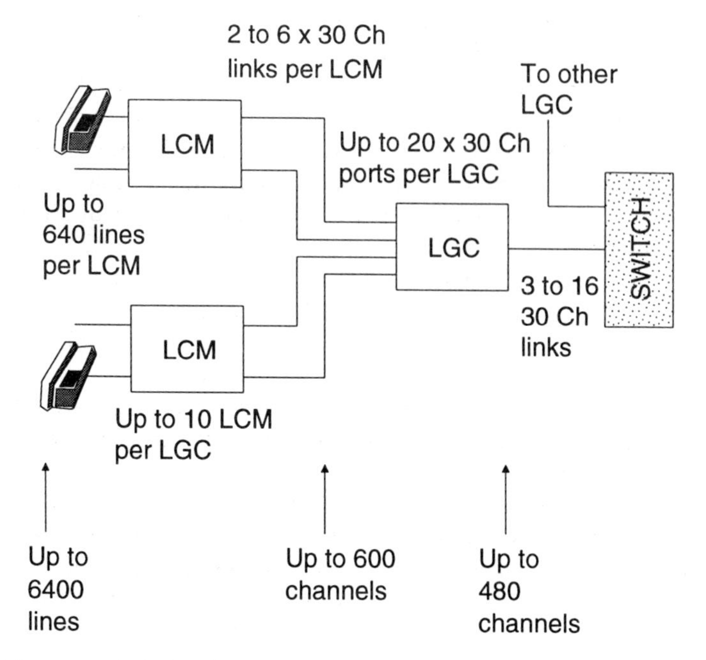

Figure 3.13 DMS line concentration

3.13.2 Digital trunk controller

This is used for 30 channel PCM connections to DMS. The PCM30 Digital Trunk Controller (PDTC) interfaces up to 16 2Mbit/s carrier systems via 16 DS30 speech links to the network (i.e. a total of 480 channels per PDTC). Its two processors operate in hot worker/standby mode.

3.13.3 Trunk Module

Trunk Modules (TM) are used for analogue trunk or junction connections to DMS. The TM encodes and multiplexes incoming speech from a maximum of 30 analogue circuits into a digital format incorporating the line supervisory and control signals. Outgoing speech and signals are demultiplexed and decoded into their respective speech and signalling information on each circuit. A range of different analogue signalling systems may be supported by selecting different interface circuit packs.

A variation on the TM is the Maintenance Trunk Module (MTM) which has all the functions of a TM card and interfaces a maximum of 28 service circuits (tones, alarms, etc.) and test equipments.

3.13.4 Link Peripheral Processor

A Link Peripheral Processor (LPP) consists of a number of Link Interface Units (LIU) which are connected to DMS-Bus via a Link Interface Module (LIM). Each LIM consists of a pair of Local Message Switches each serving a maximum of 36 LIU.

An LIU configured for No.7 signalling is known as an LIU7. Each LIU7 will provide message discrimination, routing and distribution for one No.7 signalling link. The signalling link enters the LIU7 as a 64 kbit/s (single channel) link which may have been derived from a multiplexer connected to a PDTC or directly from the PDTC as a DS0 (digital speech – no channels) link.

The LPP may be used to support Signalling Transfer Point (STP) working.

It is the architecture of the LPP which not only provides the capabilities for advanced features and services of the future, but also the ability to process the large volume of intelligent messaging required whilst maintaining the necessary firewalls to ensure total network integrity.

In first generation systems, No.7 signalling is supported by Message Switching Buffers and associated Signalling Terminals performing a similar function to the LPP.

3.14 DMS-Core

DMS SuperNode employs a distributed processing architecture. Real time functions associated with a particular facility or peripheral are delegated to local processors in the relevant sub system. A central

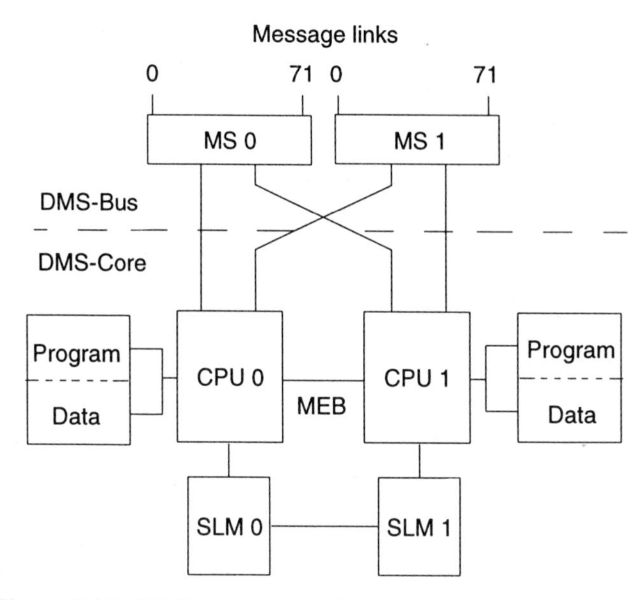

Figure 3.14 DMS central control

control (DMS-Core) co-ordinates call processing including the interactions between the Switching Network and the Peripheral Modules. (See Figure 3.14.)

There are two Central Processing Units (CPU) based on 32-bit microprocessors operating in parallel, simultaneously executing the same instructions with the same data. They are linked by a Mate Exchange Bus (MEB) for fault detection and maintenance purposes. Each has its own integrated Program and Data store, data changes in the active CPU data store being copied to the inactive one across the MEB.

Directly associated with each CPU is a System Load Module (SLM) incorporating a Winchester disk for software loads and a magnetic tape cartridge drive for archiving. This allows more rapid loading and saving of software than via the Input/Output Controller. Crossover buses are provided to allow either CPU to be loaded by either SLM. Periodic back-ups of the contents of Program and Data

Stores (Office Image) are taken on the disks and are constantly made available to the system for use in the event of a major system failure. Reloading and full recovery after a total system failure (e.g. after power supply failure) is fully automatic.

The first generation systems are generally similar but use 16-bit microprocessors, separate stores for Data and Program and are not provided with SLM.

3.15 DMS-Bus

DMS-Bus is perhaps the key architectural innovation in DMS SuperNode. It connects all system components in a uniform manner to give flexibility for future evolution.

Because all sub-systems are connected directly or indirectly to ports on DMS-Bus they can communicate freely with one another by having DMS-Bus switch messages between its input and output ports. A sub-system wishing to communicate simply creates a message and sends it in the appropriate format to DMS-Bus which then routes it to the required destination.

Assuming an average message length of 64 bytes, one fully duplicated DMS-Bus module is capable of handling about 125K messages per second.

The DMS-Bus comprises two Message Switches (MS), each having its own processor with associated memory and software. A copy of the MS load forms part of the Office Image. The two MS operate in load sharing mode but either is capable of handling all the message traffic if the other is removed from service (e.g. for maintenance or under failure conditions).

Links to the MS can be either DS30 or DS512 and messages are transmitted internally over four parallel 4Mbit/s synchronous buses.

The MS also provide the master clock facilities for the whole system.

In first generation systems, communication between the CPU and other subsystems is via Central Message Controllers, each providing up to 70 DS30 ports to the Switching Network and Input/Output Controller.

3.16 DMS software

The control component software system provides the software necessary for the basic call processing decision functions; it also provides extensive software for the administration and maintenance of the DMS-100 hardware and connecting facilities. There are several thousand software modules consisting of several million lines of program code. In order to manage the design and production of reliable software on this scale, the system has the following characteristics:

1. Programming in a high level language.
2. Programming modularity.
3. Programming structure.
4. Programming portability.

Both the Central Control CPU (first generation) and the DMS SuperNode CPU are programmed in PROTEL (PRocedure Oriented Type Enforcing Language), a high level language developed at BNR. PROTEL is a block structured, type enforcing language that enables extensive type checking on the source code at compile time.

In both generations of the DMS, multitasking in real time is based on processes that use messages to communicate with one another. These messages can be exchanged by processes within the control component; they can also be exchanged by control component processes and processes in the Switching Network or in the Peripheral Modules.

The software is structured in functional layers so that each layer provides a set of services that are available to the software layers above it. As one moves up through the hierarchy, the software provides more advanced levels of operation.

For example, at the bottom of the hierarchy is the layer which provides the basic tools required by the Support Operating System (SOS) and thus provides a set of services for all the layers above it. The top layer of the hierarchy contains the various features, such as the Traffic Operator Position System (TOPS). TOPS is an integrated operator call handling system that operates in conjunction with a DMS-100.

In addition to this functional layering, the software is also organised into a hierarchy of modules in four layers:

1. Area; a product line.
2. Subsystem; a unit of software delivery.
3. Module; a unit of loading.
4. Section; a unit of compilation.

Most of the control software created for the first generation system did not need to be modified for DMS SuperNode except to remove processor dependencies and to introduce new features to take advantage of the new processor. A proprietary Operating System (SOS) mediates between the application software and the hardware vehicle and a sophisticated compiler has been developed for re-targeting the software. Consequently, DMS SuperNode can be upgraded in future to take advantage of any new high performance processors that become available.

3.17 DMS-Link

The greater part of one of DMS SuperNode key elements, DMS-Link, is implemented in software. Essentially a software and protocol structure for use on the signalling links between DMS SuperNode applications and network nodes, DMS-Link delivers specific capabilities to end users on the telephony network to suit a variety of requirements. DMS-Link also includes a standard set of application and base interfaces for use in Custom Programming.

DMS-Link consists of sets of communicating applications and protocol stacks where the applications are software programs that provide end-to-end functions and services to the user and the protocol stacks provide the means of communication between the programs. These protocols include ISDN Access (Q.921 and Q.931) the Network Operations Protocols (NOP and X.400) X.25 for packet switched communication and Common Channel Signalling System Number 7 (SS7).

Application programs, which can be services or utilities, interface with the top layers of the protocol stacks, and by using special software interfaces, customers can add their own applications to the system as appropriate. The standard software application set currently includes call processing, OA&M processing databases and Custom Programming. Many new applications can, however, be added as desired, because of the flexibility of DMS-Link and the DMS SuperNode architecture.

The hardware for DMS-Link is based on that of DMS SuperNode and DMS SuperNode's distributed operating system controls communication between the various processors in the system. When a single application is spread across multiple processors, the operating system allows the various elements of that application to communicate. When several similar functions are integrated across a number of switching nodes, they are said to be networking.

Networking with DMS-Link provides a seamless view of a function, whether it is the transportation of the calling number from originating to terminating node, or the transmission of a traffic summary for a given Centrex customer to multiple nodes. This seamlessness is achieved through DMS-Link's use of protocols (comparable to languages) and applications (comparable to semantics) which together help create a suitable environment for 'conversions' between nodal functions. This environment, which disassociates the customer's logical network from any specific physical implementation, creates a platform for launching new services.

3.18 DMS OA&M

The maintenance and administration area consists of all the Input/Output Devices and their controllers used to support the operation and maintenance of the switch. These may include a selection of disk drives, tape drives, printers, terminals, modems, PCs and data link controllers. They may be interfaced onto the DMS-Bus either directly via Input Output Controllers (IOC) or via Ethernet Interface Units supporting Local Area Networks (LANs). A minimum of two IOC are always provided to guarantee terminal and modem access to the system at all times.

The maintenance and administration workstations on DMS switches are known as MAP, Maintenance and Administration Positions. These consist of a terminal VDU and keyboard (the man machine interface), audio visual alarm panel (providing Minor, Major and Critical alarm indications), headset or keyset telephone system (for voice circuit monitoring) and integrated test jacks (for using non-standard external test equipment).

The MAP may be used, subject to terminal and user security screening, to access a variety of maintenance and administration software environments. These include:

1. General exchange maintenance.
2. Network management.
3. Service analysis.
4. Operational and traffic measurements.
5. Trunk testing.
6. Line testing.
7. Data modification and service orders.

Full screen displays are used to provide a 'live' picture of overall system and subsystem status as well as menu driven command prompting for maintenance, testing and data modification.

IOC are interfaced onto the DMS-Bus via two DS30 message links (one to each Message Switch). Each IOC consists of a firmware Message Processor Card and up to nine Device Controller cards, each with four ports.

Devices connected to Local Area Networks (LANs) may access the DMS-100 system through Ethernet Interface Units (EIUs). These are housed in peripheral modules called Link Peripheral Processors.

3.19 Further evolution with DMS

DMS SuperNode architecture is now not only providing a firm foundation for other switching products to build on, but its enhanced functionality is making advanced features and sophisticated services happen. Furthermore, the Intelligent Network is one step nearer since common channel signalling started using DMS SuperNode based

Signal Transfer Points to bring to the domestic subscriber a type of service previously only available on large private business exchanges.

4. PABX and key systems

4.1 Purpose of key systems and PABXs

Prior to the existence of key systems and PABXs (also referred to as PBX) a company would have a separate telephone line from the public exchange for each employee requiring a telephone. If a call came in to the wrong telephone then either the caller would have to call back to the correct telephone or the correct person would have to physically come to the called telephone. Also, the company would have to pay the rental charge for every telephone line used.

With a key system or PABX, incoming calls can be answered by one person and then the call transferred to the correct person. The number of telephone lines from the public exchange is now determined by the expected maximum number of simultaneous calls (i.e. traffic) rather than the required number of telephones. Key systems and PABXs pay for themselves by the reduction in exchange line rental costs, the increased convenience to callers, and the increased productivity of employees (Thomas, 1991; Prentice, 1994; Armstrong, 1995).

Early key systems and PABXs were distinguished mainly by how they answered and made external calls. Key systems were (and still are) mainly used in smaller locations without a centralised answering position. Incoming calls ring several telephones and employees have a collective responsibility to answer calls and transfer calls to the correct person. PABXs usually do have a centralised answering position where the prime responsibility is to answer incoming calls and transfer them to the appropriate person. Key systems usually have a visual appearance for each exchange line showing whether it is used or not. To make an outgoing call, a free exchange line is manually selected and the required number dialled. PABXs usually pool exchange lines together and require an access code to be dialled to seize an outgoing exchange line. After dialling the access code, the receipt

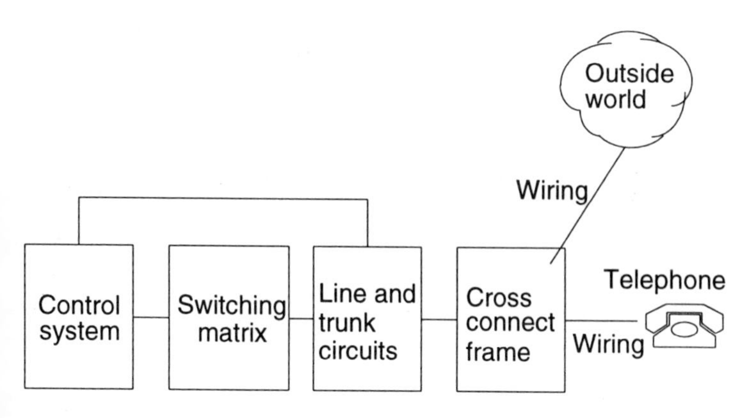

Figure 4.1 Main elements of a PABX or key system

of dial tone indicates that an exchange line has been successfully seized while busy tone indicates that there are no free exchange lines at that time.

In recent years, the differences between key systems and PABXs have become blurred. Key systems now offer centralised answering positions and access codes to exchange lines while PABXs offer many features traditionally only available on key systems. Many manufacturers now refer to their products as hybrid PBX key systems.

Figure 4.1 shows the main elements of a PABX or key system.

4.1.1 Control system

In very early equipment, the number of services supported were very small. There would be an ability to recognise off-hook states, provide dial tone, recognise dialled digits, route calls, and ring telephones, but not much else. The control system was more of a concept than a physical reality. Each part of the system controlled the functionality of that part of the system. In modern day terms, this would be referred to as distributed processing.

As technology changed, the architecture changed to consolidate the major control elements in one part of the system. The technology used in this changed from relays to wired logic to microprocessors, but the fundamental principle has remained the same.

As microprocessor technology has evolved and become more cost effective, there is now a trend to move back towards more of a distributed processing environment. Processors in various parts of the overall system off load tasks from the central processing unit, and hence improve the overall performance of the system. Individual products have different architectures in terms of what tasks to perform locally and what tasks to perform centrally, but the fundamental principle of a hierarchical structure of processors communicating with each other, and with the rest of the system, is now common.

4.1.2 Switching matrix

This is the part of the system which connects one peripheral device to another. In step by step (Strowger) technology, the switching matrix consisted of a series of line finders, group selectors and final selectors which, under the control of the dialled digits, routed a call to its final destination (see Chapter 1). Physical movement caused one pair of wires to be connected to one of a number of other pairs of wires depending on the destination of the call. The number of elements (e.g. group selectors) in the matrix was determined by the expected number of simultaneous calls (traffic) through that part of the matrix. Higher than expected traffic levels would cause calls to be blocked and hence not reach their final destination. Since the call was routed through the matrix in a step by step manner under the control of the dialled digits, there was no opportunity to try alternate routes through the matrix.

Crossbar technology (Figure 4.2) used relays to make connections between wires in a physical matrix. Rather than responding to individual digits, all the required digits would be sent to the control system which would then find a possible route through the switching matrix. While calculating the required number of matrix elements was still required, this method of operation was more efficient since all possible paths through the matrix could be tried. The other major advantage of this method was that routeing information could be different to the dialled number information. This made it much easier

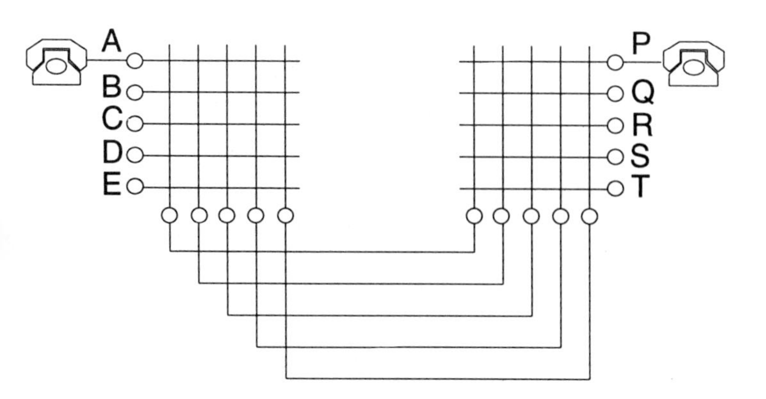

Figure 4.2 Simple crossbar matrix

to achieve the goal of one telephone number regardless of where the call originated from.

Silicon controlled rectifier (thyristor) technology simply replaced the crossbar matrix with an electronic equivalent. However the cost effectiveness of this technology allowed non blocking matrices to be produced, thus reducing, or even eliminating, the need for matrix engineering.

All the above switching matrices are referred to as analogue systems. That is, once the human voice is converted into an analogue electrical waveform, it remains in that form until converted back to sound in the telephone receiver.

The next major technology change was to digital systems. In a digital system, the analogue electrical waveform is sampled, and the samples converted into a digital code. At the far end of the call, this digital code is converted back into an analogue sample. The series of samples are then filtered to convert back to an analogue waveform. The most common digital encoding scheme is called Pulse Code Modulation (PCM). In PCM, the analogue waveform is sampled 8000 times a second, and each sample is converted into an 8 bit digital code. The bandwidth of PCM is therefore 64000 bits per second (64kbit/s).

Digital systems use Time Division Multiplex (TDM) to interleave digitally encoded telephone calls onto a common highway. The management of this highway (or highways) in effect creates a switching matrix.

4.1.3 Cross connect frame

The control system and switching matrix are all normally located in one part of the customer's building. The telephone instruments on the other hand are distributed throughout the customer's building. The cross connect frame is therefore used to connect pairs of wires from individual telephones to individual circuits in the system peripheral equipment. Similarly, the cross connect frame is used to connect circuits from the outside world to the peripheral equipment.

4.1.4 Telephones

Telephones perform a number of fundamental purposes. On receipt of a signal from the telephone system indicating an incoming call (ringing), the telephone is expected to deliver an audible and/or visual signal to alert the called person. When the called person answers the call (goes off-hook), the telephone must signal the telephone system to remove the ringing signal and cut through the transmission path. The receiver in the telephone converts electrical energy representing the caller's voice into acoustical energy. The transmitter in the telephone converts acoustical energy into electrical energy. A part of the telephone circuitry sends some of the electrical energy from the transmitter to the receiver so that the user can hear their own voice. This is called sidetone (see Chapter 9). When the called person disconnects the call (goes on-hook), the telephone must signal the telephone system to this effect.

On an outgoing call, the telephone must request service from the telephone system (by going off-hook). The telephone system usually returns dial tone to the caller at this point. The telephone must now indicate to the telephone system the number of the called person. Early telephones disconnected and reconnected the circuit to the telephone system to send pulses (dial pulses) to indicate the number dialled. More modern telephones send bursts of tones (DTMF) to achieve the same end. The transmitter and receiver perform the same functions as for incoming calls, and a disconnect signal is sent at the end of the call.

PABXs and key systems require the telephone to perform more than the above fundamental functions.

In order to perform functions such as transfer and conference, a recall signal must be sent from the telephone to the telephone system. Two types of recall signal are in common use. The first creates a temporary disconnect in the circuit, similar to a dial pulse. The second connects an earth (ground) to the circuit.

For PABX and key systems operation, the technology available on standard dial pulse and DTMF telephones became a limiting factor. Special telephones were developed for use with particular telephone systems. To begin with, multi-pair cables were used between these telephones and the associated telephone system to transmit the various signals required to perform additional functions. Later, electronic circuitry was used to transmit these various signals over a single pair of wires using some kind of digital code. At that time, the electrical energy representing the person's voice was still transmitted between the telephone and telephone system in an analogue format, and hence these telephones were still classified as analogue telephones despite the fact that the signalling was now in a digital format. The next stage in telephone set evolution was to move the conversion from analogue to digital voice (codec) from the telephone system to the telephone itself. Telephones with a codec built into them are classified as digital telephones.

Using the techniques of Time Compression Multiplex (TCM), a digital telephone is able to multiplex the digitalised voice and digital signals onto a single pair of wires.

4.2 PABX architecture

Figure 4.3 shows a more detailed block diagram of a PABX. The trunk circuits provide the basic access to and from the public network and can be analogue or digital connections. A variety of signalling systems are in operation ranging from basic loop disconnect signall-

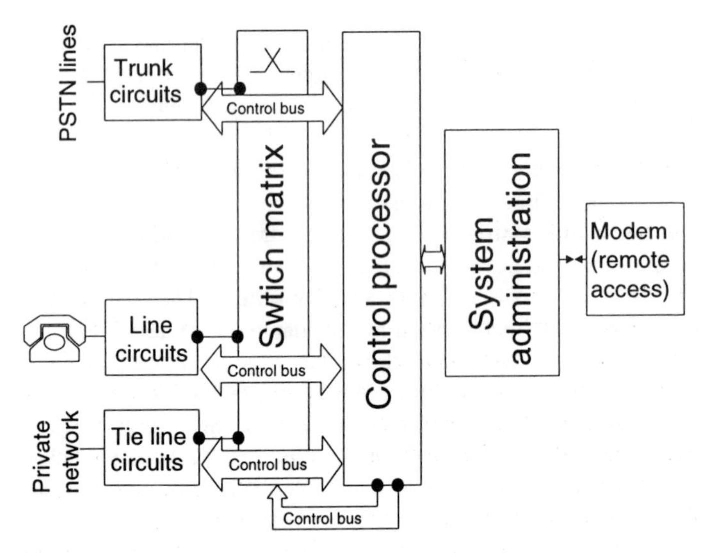

Figure 4.3 PABX block diagram

ing to common channel signalling on a 64kbit/s channel of a 2Mbit/s 30 channel trunk circuit.

Tie line circuits provide access to the private network interconnecting a number of PABXs directly or indirectly via tandem private exchanges. The signalling on these circuits range from basic E&M through the spectrum to common channel signalling carrying DPNSS or Qsig signalling, with feature rich inter-PABX functions transferred over the signalling channels during call set up and during the conversation phase of the call. Line circuits provide the basic BORSCHT functions for analogue telephones.

Special proprietary analogue line circuits are in use for multifunctional terminals and digital multifunctional telephones, with data transmission on either the D-channel or the 2nd B-channel. The switching matrix can be either CMOS analogue switches for the smaller systems and digital switching is the most cost effective on nearly all other designs. The system control provides all the operational software/control functions to interconnect the various functions.

4.3 Trunk circuits

Trunk circuits provide the interface to the PSTN and use a number of different signalling systems to interact with the network.

4.3.1 Loop calling

This form of trunk circuit is the simplest interface which mimics that of the basic telephone interface to the public network (see Chapter 2). The circuit is a 2-wire connection to the public exchange with the basic elements shown in Figure 4.4. In the idle condition the public exchange presents a negative d.c. voltage on one wire and either an exchange earth on the other or the positive side of the exchange battery on the other wire, depending on whether the exchange has a floating battery or a battery connected to ground.

The PABX circuit presents the off-hook impedance of the ring detect circuit to the exchange. Incoming calls from the exchange to

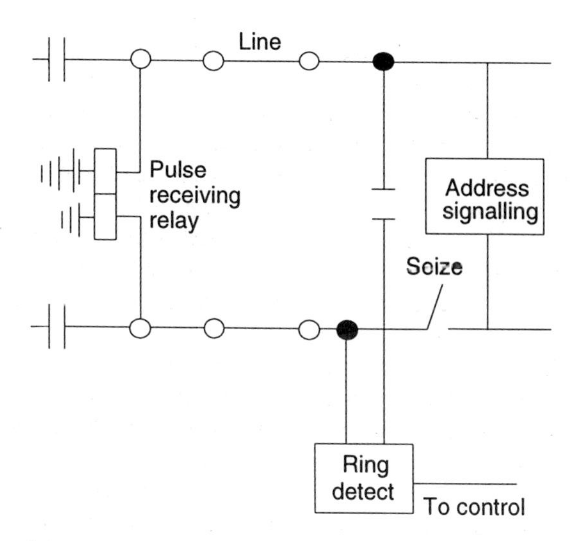

Figure 4.4 Loop calling

the PABX are ringing current superimposed on the d.c. of the exchange battery. The capacitor in the ringing detector circuit blocks the d.c. exchange battery and the a.c. ringing signal is presented to the ring detector circuit, which generally consists of an opto coupler which isolates the line from the low level electronic circuits. The a.c. ringing activates the opto coupler and the signal is presented to the system. It is normally the system software which performs the persistency checking on the signal and timers are set to monitor the on and off periods of the ringing signals.

As no common international standards exist for ringing signal cadences, different public exchanges provide different ringing signals depending on the individual operator's specification, so the PABX circuit must be flexible enough to cater for different requirements.

The problem that can arise with loop calling lines is call collision due to the line seizure being applied by the public exchange before the ringing signal is available at the PABX. This condition arises from the nature of ringing signals as they are bursts of a.c. signals followed by a silent period. It is during this silent period that the exchange line may be seized simultaneously by the public exchange and the PABX trunk circuit.

To initiate a call the PABX trunk circuit will seize the exchange circuit by closing a contact and looping the line. The public exchange will respond by returning dial tone superimposed on the d.c. exchange battery feed within a specified period of time depending upon the network specification. At this point the PABX can go into the addressing phase which can be either loop disconnect signals at 10pps or DTMF signals. In either case the out pulsing of digits is controlled not by the user's telephone but by the system software of the PABX so that features such as short code dialling etc. can be implemented in the PABX. The address signalling is therefore regenerative signalling and PABX can have 10 pps, DTMF, or proprietary signalling from the PABX telephones to the PABX and then the address signalling is regenerated as required towards the public exchange.

A call can be cleared by either the distant end or the PABX removing the loop from the line giving a disconnect clear. The main problem for the PABX is knowing when the distant end has cleared as the disconnect clear signal is not normally transmitted from one end of the network to the other and in some cases dial tone may be returned to the PABX while the PABX still has the line seized. The system relies on both parties clearing. For this reason, and the call collision mentioned earlier, high traffic PABX trunk circuits use earth calling or ground start signalling.

4.3.2 Earth calling

Earth calling signalling (also known as ground start) was introduced when automatic direct outward dialling from the PABX extensions was introduced and the telephone operator was removed from monitoring the progress of the call and hence the disconnect signal. The major reason for the introduction of this signalling system was to prevent call collision during the silent period of the ringing signal.

The principle of operation of the earth calling signalling is shown in Figure 4.5. To request service the PABX applies local earth via a relay contact to one leg of the pair from the public exchange. The exchange detects current flowing in this wire and responds by applying the exchange battery across the pair with dial tone superim-

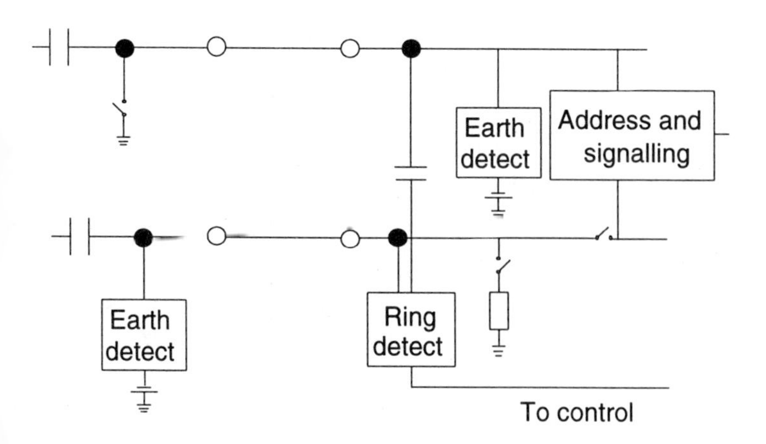

Figure 4.5 Earth calling

posed. The PABX can then go into the addressing state like loop calling and send the digits to the exchange in either 10pps loop disconnect or DTMF signals. On an incoming call to the PABX the public exchange applies an earth potential to one leg of the pair (the opposite leg to that used by the PABX to call the exchange) and the PABX detects this signal as a calling signal.

Ringing is also applied by the network but is now of less importance and the PABX does not need to wait for the ringing signal during the silent period before acting on the incoming call signal. Either end can disconnect the call by breaking the loop, but the PABX is required to run a timeout and maintain the PABX trunk busy to all other PABX users until the public exchange has had time to clear the call down and return the line to its idle condition.

4.3.3 Meter pulse detection

On any of the 2-wire analogue signalling systems the PABX may also be required to detect meter pulses sent from the exchange. These are normally only used on systems where it is necessary to provide accurate call accounting within the PABX e.g. in a hotel application or where the PABX is in a multi-tenant environment and the PABX owner charges for the use of the system on a timed basis related to the meter pulses received from the public exchange.

Two signalling systems are in use for meter pulse detection and both use out of band signalling as it is necessary to send the meter pulses to the PABX during the conversation phase of the call and the talkers should not be able to detect the pulses. 50Hz longitudinal pulses are applied to the speech pair and the trunk circuit of the PABX must therefore be equipped with a detector which senses the meter pulses and then passes them onto the system controller for assimilation with the call record. The other method uses 12kHz or 16kHz signals.

4.3.4 2Mbit/s digital access

2Mbit/s digital access is presented to the PABX as a coaxial circuit carrying 30 channels for voice or data, a synchronisation channel on timeslot 0 and a signalling channel on timeslot 16. In the UK DASS2 is the signalling system employed on the primary rate ISDN service. The signalling system is based on the ISO 7 layer reference model with Layer 1 using a ITU-T G.703 2.048Mbit/s interface utilising HDB3 line coding. The PABX must be designated as the slave as far as synchronising to the network is concerned and this is achieved via timeslot 0.

The messaging between the PABX and the public exchange is via timeslot 16 using Link Access Protocol D (LAPD) which is designed for end-to-end signalling over ISDN and is an extension of the ISO High Level Data Link (HDLC) control procedure.

LAPD is a frame orientated protocol with the information it carries consisting of messages that are delimited by flags. Error correction of information frames is by re-transmission until correct or until a timeout expires. The next frame is not transmitted until the last is correctly acknowledged.

Within DASS2 there are three types of call identified:

1. Category 1 calls, which are calls that require end-to-end digital paths e.g. a data call. In this type of call no in band tones or announcements are given by the network.
2. Category 2 calls, which are calls for which an end-to-end digital path is not essential and if a non-digital path is encountered then a 'not suitable for data' message is sent back to the PABX.
3. Telephony, which cannot be used for data and the swap facility in the system cannot be used, nor can the user to-user supplementary data services.

The public network offers 5 services to the DASS2 PABX user as follows:

1. User to user signalling, which allows several blocks of 32 bytes of data to be transmitted between the calling and the called PABXs. There is also the option to configure one or more DASS2 speech paths as tie lines, to permit transmission of large amounts of data end-to-end from user to user.

2. Closed user group facility enables a logical grouping of users within the public network. Connected party identities can be displayed on the PABX feature phones. Each group is identified by a number and the network places restrictions upon calls into and out of the group. Most of the functionality is provided by the network and the PABX has to provide the appropriate call set up and acknowledgement messages.
3. Calling line identification provides the PABX with the identity of the calling customer and apart from displaying the calling line number on the operator's position on a PABX, it can also be made use of in ACD applications where the customer's line number can be used as an additional check when accessing a data base via a CSTA link.
4. Call charge indication is an attractive feature for PABX applications in hotels. It provides information on customer originated calls and, in the basic PABX, an input to the call management packages. At the end of each call the network reports the number of units used and the cost of the call.
5. Network address extension is only available on category 1 calls and enables a destination code to be appended to the PSTN number. The feature can be used to access a specific device on a LAN connected to the PABX.

4.4 Private networking

Many PABXs are interconnected by private networks carrying both voice and data traffic. Access to these networks is provided by tie line circuits in the PABX which provide the signalling and interconnection between the PABX and the transmission network of the private system. Some private networks are not point to point and the PABX must then act as a tandem exchange interconnecting the tie line circuits and providing a digit translation facility, to enable the user to dial the least number of digits to get from one point to the other.

With the deregulation of telecommunication services the PABX operator is able to offer 'break out' facilities over the private network. When the user dials a PSTN number, the PABX will interpret this number. If it is more economical to send it over the private network.

the call will be routed to the nearest geographical point in the private network to the required PSTN number, normally within the local call charging area. It will then 'break out' into the public network via a normal PSTN trunk circuit.

These types of facilities require intelligent tie line signalling systems and like the PSTN signalling, tie line signalling comes in varying degrees of complexity depending on the application within the private network.

4.4.1 E&M signalling

E&M signalling is the simplest form of tie line signalling and has been borrowed for PABX usage from the public network where it has been used to interface public switching systems with the long distance transmission line signalling systems. The signalling scheme can be applied to both 2-wire transmission circuits and 4-wire transmission circuits as the signalling is carried on an additional two wires between the PABX and the distant end, or the intermediate transmission or multiplexing equipment.

The E&M signalling system derives its name from the historical designations of the signalling leads between the PSTN trunk circuits and the transmission equipment which tended to be housed in different parts of the building of a public exchange. The E&M signalling circuits use one lead for each direction of transmission and a common earth return which makes them more prone to noise than a balanced 2-wire system.

The E-lead has an earth potential applied from a contact and the receiving equipment has a battery connected relay as the receiving circuit. The same circuit is used in the opposite direction on the M-lead but in reverse, as in Figure 4.6. The impedance between sending and receiving ends is thus limited to about 140 ohms.

In the UK the E&M signalling system is called SSDC5 and is now only used between equipment in the same building. The relay contact is pulsed at 10pps from the PABX control system to send digits into the private network. It is possible to use the E&M circuit only for line seizure and release signals and use MF inter register signalling for address signalling.

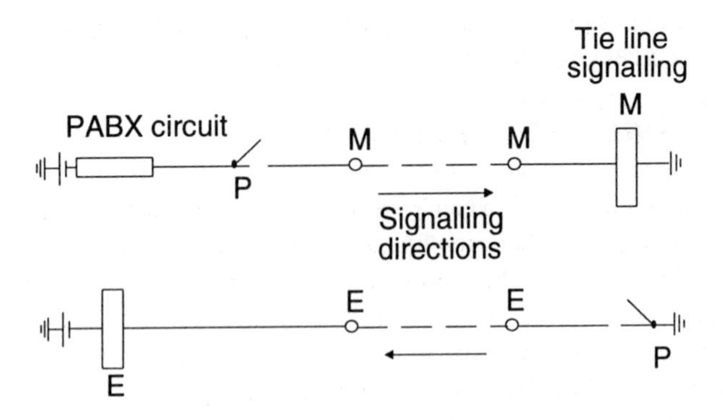

Figure 4.6 E&M signalling

4.4.2 Double current signalling

With double current signalling a polarised relay is used which is very sensitive and the signals are transmitted as opposite directions of current flow as in Figure 4.7. The use of double current signalling also minimises the effect of line capacitance by continually reversing the current in the line. The limitation of double current signalling is that

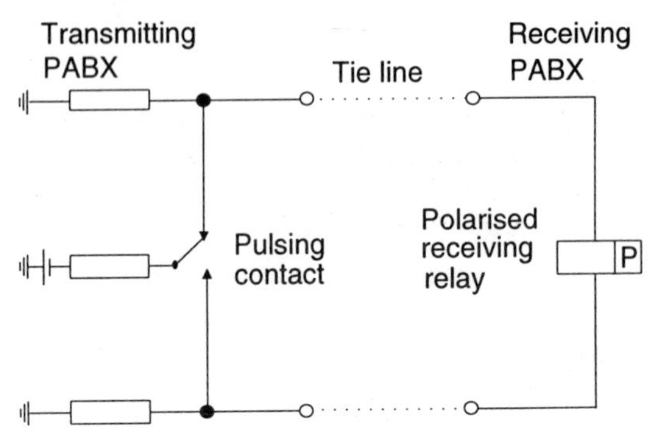

Figure 4.7 Double current signalling

the address signalling can only be sent at 10pps and a physical pair is required between the PABXs.

4.4.3 Single frequency signalling

Single frequency a.c. signalling has been developed for use on 4-wire amplified circuits between PABXs enabling line and address signalling to be sent over the amplified circuits on unlimited distances for the PABX user. The signalling system was derived from the signalling circuits used in the 4-wire amplified circuits between the public exchange and the long distance transmission systems.

An important requirement was the choice of tone for line signalling to avoid imitation of the signalling tone by the voice signal. This was achieved by selecting a tone high enough in the speech band such that the power within the voice signal was at its lowest level. The tone selected was 2280Hz.

In the UK the signalling system is SSAC15 and is similar to CEPT L1 which makes it suitable for international use. The system operates on a forward and backward signalling basis, as in Figure 4.8, with

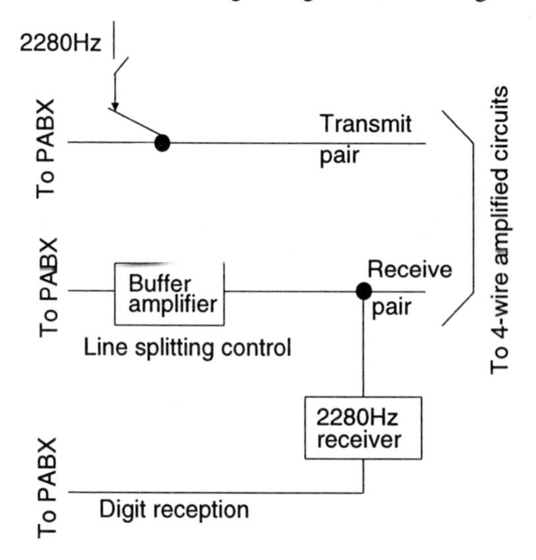

Figure 4.8 Single frequency signalling

both directions of signalling separated. Signals are sent as either the application of the 2280Hz frequency (tone on) or removal of the frequency (tone off).

The system operates on a link by link basis and signals are contained within each link. Forward and backward signals may be passed simultaneously and in some cases when speech is present but the primary use is for call set up and clear down. The system allows for address signals to be sent at 10pps.

Like E&M signalling on physical wires this signalling system allows the use of multifrequency inter-register signalling for address signalling and the 2280Hz for the line signalling.

4.4.4 Inter-register signalling

Inter-register signalling, like its counterpart in the public network, is intended to provide the PABX private network user with fast call set up and the use of a set of supplementary services across the network, similar to those available between the the telephone users on the same PABX.

The signalling system is based on ITU-T R2 signalling using multifrequency signalling on either 2-wire or 4-wire circuits between PABXs. The signalling system requires a supporting line signalling system on each link, such as E&M or SSAC15, to control seizing, answering and clearing signals.

The signalling system operates on an end to end basis with the signalling path being progressively extended as the call progresses through the private network tandem switches. It is a compelled system i.e. a forward signal once sent remains on the line until a response is received from the distant end. Once the forward removal signal has been detected at the distant end this causes the removal of the acknowledgement signal and completes the compelled signalling cycle.

Twelve frequencies are used, divided into upper and lower bands for signalling in the forward and reverse direction respectively. Each frequency is spaced 120Hz apart starting at 540Hz and going through the range to 1980Hz. The frequencies are sent as 2 out of 6 combinations giving 15 discrete combinations in each direction.

The system is further extended by reserving some of the combinations as shift functions so that the basic combinations can have up to 3 meanings. The later is used to add features between PABXs such as three party conference, executive intrusion, call offering etc.

The main use within PABX private network operation is to provide fast call set up and the basic set of signals is defined for the sending of routing digits between registers.

4.4.5 Dual Tone Multi Frequency

As PABXs were modernised and took advantage of DTMF telephones on the extensions, the need for fast call set up over private circuits became very apparent and the post dialling delays on tie lines using 10pps E&M or AC15 signals was not adequate for the modern business needs. R2 signalling was used to overcome this problem but only the larger PABX systems could afford the cost of implementing the full R2 system with its forward and backward signalling. The smaller PABX manufacturer found that DTMF receivers had to be provided for interpreting the address signals from DTMF extension telephones. When public exchange DTMF lines became available DTMF senders also had to be provided for the regeneration of signals and the pulsing out of short code dialling stores. This meant that the PABX had all the elements within the system for fast call set up using DTMF signalling over the private network and it was merely a software task to reorganise the PABX to make better use of these existing facilities.

Fast call set up over AC15 tie lines was then possible by using the AC15 for line signalling seize, release, and answer. The AC15 10pps address signalling was then suppressed and the sending PABX connected a DTMF sender to the tie line; the distant end on seizure connected a DTMF receiver to detect the incoming DTMF signals.

On the simpler PABXs where no restrictions were placed on dialling or where the routing digits required by the private network were minimal, the PABX user's telephone was directly connected to the tie line and the DTMF pulses dialled directly into the distant PABX. The frequencies used in this form of signalling are identical to those of the basic telephone.

4.4.6 Digital private network signalling

In the early 1980s 2Mbit/s private circuits became available for use in private networks and this led the need for an inter-PABX signalling system that would maximise the benefits of the all digital network. In the absence of any international standards developments in this area UK manufacturers and BT formed a working party to create an open standard so that PABXs from different manufacturers could interwork freely within a network for at least the simple call set up process, but also capable of implementation of advanced features as the need arose in the customer's network. Not all PABXs in the network needed to be capable of implementing the advanced features such was the flexibility built into the specification.

Digital Private Network Signalling System No 1 was first published by the working party in May 1983 and has since been revised and extended as customers' needs grew. DPNSS is a common channel signalling system on 2Mbit/s links using ITU-T G.703 interface with the signalling information carried on a common 64Kbit/s channel in timeslot 16 for the 30 voice or data channels. Synchronisation is provided by the one remaining channel in timeslot 0.

DPNSS extends the facilities normally only available between extensions on a single PABX to all extensions on all PABXs that are connected together in the DPNSS private network. It specifies the signalling for a wide range of facilities for voice and circuit switched data calls.

DPNSS is organised on the ISO 7 layer reference model with the Physical Layer 1 and the Link Access Layer 2 virtually identical to the DASS2 equivalent layers. Layer 3 provides the basic messaging for DPNSS between the PABXs and these messages are implemented entirely in the software of the PABX.

Messages can be transmitted transparently through a PABX in the network if the destination address points to a PABX not linked directly to the PABX originating the call. When a call is set up on DPNSS the originating PABX sends an initial service request message (ISRM) with information on who the caller is (OLI), category of the calling line (CLC), a service indicator code (SIC) and the destination address (DA). This is transmitted over timeslot 16 to the distant

PABX which responds with a number acknowledgement message (NAM), the called line category (CLC) and the called line identity (CLI) if the called line is free. If the called line is busy the distant PABX responds with a clear request message and busy indication. (See Figure 4.9.)

Channel 16 is divided to permit virtual and real messages to be transferred, where the real messages are used to set up calls for normal voice activity and the virtual channels can be used to pass supplementary information to enhance the interconnection of the PABXs. Within the PABX network DPNSS has had to define specific functions of a PABX. A transit function interconnects a call between two DPNSS channels. The call may be the initial set up via a new channel or a supplementary service call over an existing channel.

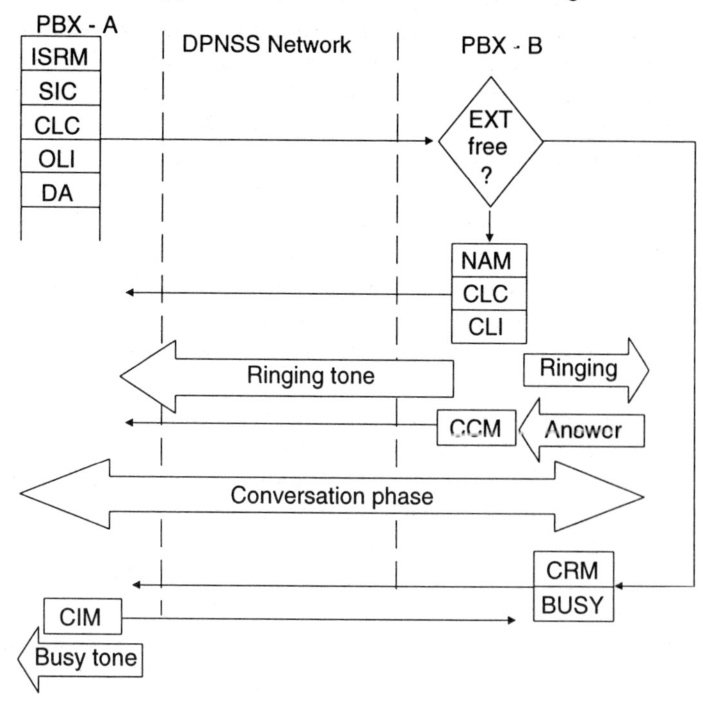

Figure 4.9 DPNSS call set up messages

An end function interconnects a DPNSS channel to an extension telephone, a non voice terminal, a route to another signalling system or a conference bridge. The end function may be provided by an originating PABX, a gateway to another signalling system, a gateway from another signalling system, a terminating PABX, or a conference PABX. A branching function may occur when some supplementary services result in a new (related) call sharing a channel on part of its path from the originating PABX with the existing call. Branching functions may occur at the originating, terminating, transit, or gateways. As far as each call is concerned the branching function is always accompanied by either an END or a Transit function.

The DPNSS specification is divided into a number of sections with 1 through to 6 covering the basic call set up and mandatory interworking requirements. Sections 7 to 39 define supplementary services and it is these that make the networking of PABXs a very powerful, feature rich system.

4.4.7 Analogue private network signalling

As the PABX private networks are places of high innovation and largely free from bureaucratic restrictions the users have found a need to re-use their existing analogue tie line private circuits, but with enhanced features that came with DPNSS. Providing 2Mbit/s links to all PABXs on a network with 30 voice or data channels was an overkill for all applications, particularly if the network had a number of branch offices with a small number of tie lines connecting them to the main network. If a user wanted the same features available at the small branch office as was available on the main network then a signalling system as comprehensive as DPNSS but which would operate over existing tie lines would be required.

This problem was solved by PABX manufacturers by re-using all the software that had been developed for DPNSS and redirecting this over a data link between the two PABXs. Each PABX is equipped with an asynchronous signalling circuit which converts the PABX processor bus into a serial signalling output that can be connected to a modem. The modem is used to convert the digital signals from the PABX into analogue signals to run on the analogue tie line between the PABXs. It is selected to run at a low speed 1200, 2400 baud etc. depending upon the private circuit available between the two PABXs and the expected message traffic for the number of speech channels between the PABXs.

The voice traffic is carried on separate private circuit tie lines that do not have any signalling system and are kept open at all times. Availability of circuits and feature transfer is controlled on the inter-PABX signalling channel via the modems with messages identical to those used in DPNSS.

4.5 Line circuits

4.5.1 Analogue line circuits

Analogue PABX line circuits are very similar to those of public exchanges providing the same functionality for connection of the plain ordinary telephone.

The PABX line circuit needs to provide the BORSCHT functions but the specification can be less severe in some areas, as the line length of the average PABX line is under 1km.

4.5.2 Key system line circuits

In key telephone systems the line circuits differ from that of a PABX in that additional signalling information needs to be transmitted to and from the telephone to provide the functionality of a key system. No standards exist for these signalling systems and they are all proprietary to the individual key system manufacturer. However the principle is the same in all the systems even if the physical and logical implementations may differ.

The key telephone and the line circuit have a bi-directional data circuit that connects the processor in the telephone with the processor in the key system. A message is sent from the the key telephone to the system requesting service and this is acknowledged by the system returning a message to light a lamp or change the indication on the telephone display. Messages are sent between the telephone and the system until the call is completed. The system may require to send messages to idle or busy telephones on the system, depending on the system configuration, informing each telephone of the status of all the other telephones, the status of tie lines, or trunk lines. As the system size increases so does the need for more messages on the line.

In this form of system the telephone is normally analogue and similar to an ordinary telephone but the signalling can be digital or analogue on a separate pair of wires from the system to the telephone, depending upon the individual designs.

4.5.3 Digital line circuits

Digital line circuits take many forms depending upon the transmission system employed between the telephone and the line circuit and within the PABX and key system environment many of the systems are proprietary, being designed for the most cost effective arrangement. If the system is 4-wire then one pair will be used for transmit and one for receive and the circuits at either end are connected with transformers. ITU-T has now defined a system that works on a 4-wire basis between the terminal and the line circuit (in ITU-T I.420/430) which operates to the S reference point standard at 144kbit/s with 2B+D channels.

In 2-wire systems the line interface is more complex and two line systems are common, burst mode or echo cancellation. Echo cancellation is much more complex to implement in silicon and PABX line circuits have tended to use the simpler burst mode technique. Echo cancelling is a full duplex solution in that it transmits data continuously in both directions at the same time. In burst mode operation the line circuit transmits a burst of data to the terminal, the line is switched off for a guard period and then a burst of data is transmitted from the terminal to the line circuit. The line transmission rate in burst mode systems is therefore much higher than in echo cancellation systems for the same rate of information transfer between the terminal and the line circuit. The technique is based on the fact that duplex transmission can be achieved on a single pair of wires by increasing the line transmission rate by some factor greater than 2 such that the

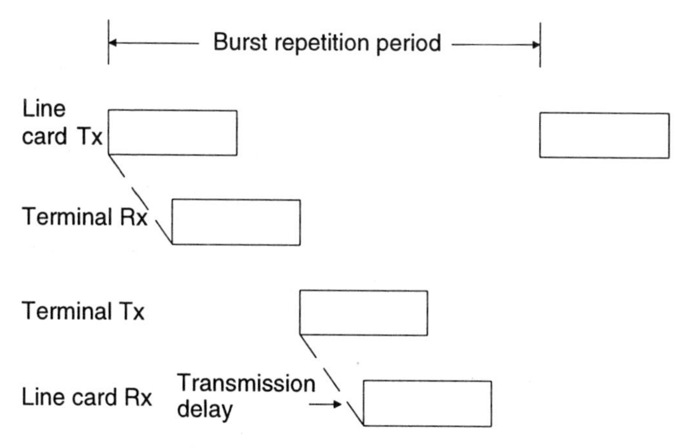

Figure 4.10 Burst mode transmission

transmit and receive signals at each end are compressed in time and never overlap, as in Figure 4.10.

Transmission rates of 384kbit/s on cable up to 1000 ohm loops are possible. A typical transmission system would have the following key parameters: 4kHz burst repetition rate; 2 PCM characters (16 bits); 1 data word (16 bits); signal/aux channel (4 bits); total bits/burst (36 bits); 384kbit/s line transmission; scrambled AMI pulse transmission.

Figure 4.11 shows how the voice and data are combined on the one pair of wires, then separated at the line circuit end to be switched through the system.

4.6 Control and administration

In the modern PABX the system is under software control, running on either a single processor for the smaller systems or with the functions divided between processors in the larger systems. In any case the particular architecture is governed by the system designer's needs to balance cost against the functionality specified. In most systems, like any computer system, the software runs on a processor(s) with an operating system handling the basic low level functions and then the application layer interacting with the operating system.

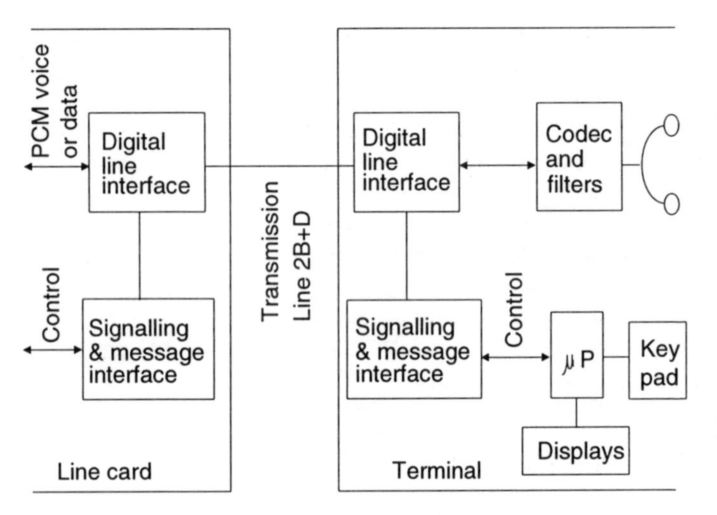

Figure 4.11 Voice and data over one wire pair

All the call control software operates in real time and only peripheral facilities, like call statistics and database administration, operate in non real time mode. Control of a PABX is all centred round the software, which can be visualised as a number of concentric rings with the real time multitasking operating system in the centre surrounded by the hardware drivers e.g. the Layers 1 and 2 functions of the OSI 7 layer model. The database for the system is then the application software. It is in the application software that all the logical functionality of the switch resides and this is the most dynamic area of the software, which is required to evolve with the users' needs.

Access to the database is of prime importance to the PABX user, as this needs to be regularly updated to keep pace with the changing needs of the business. Early designs of PABX software were complicated to use and relied on the system manager understanding not only the business requirements but also how the PABX database bit pattern was organised, since changes to the database involved laboriously entering bit patterns at the appropriate points to re-programme the system. This led to the development, particularly in the US, of off line

PC controlled database management systems, which had a more friendly user interface. This was used to program the new requirements off line and then up load this onto the PABX during a quiet traffic period. PABX suppliers have in the main now developed menu driven database management software that is resident as a set of application programs in the PABX and allows online changes to be made to the database without the need for out of hours working by the system manager. The application software is the most complex and handles a number of functional areas such as telephone or terminal interface, incoming call disposition, outgoing call handling trunk or tie line, database administration, system statistics and interfacing to other peripheral equipment.

4.6.1 Telephone or terminal software

Telephone or terminal software interprets the commands and functions invoked from the telephone or multifunctional terminal. Apart from receiving the dialled digits validating them, and passing them onto the outgoing call handling software, the telephone or terminal handling software controls all the user features.

On a basic PABX telephone, with only DTMF signalling, access to features is restricted to dialling special codes, which usually means that any call in progress needs to be put on hold before a feature such as conference another party, or dial ahead to determine if a colleague is free or busy. This limited signalling system and the absence of visual feedback for confirmation of feature invocation, has led to the development of multifunctional terminals with displays and arrays of keys to enable easy feature activation.

Typical features on the modern PABX would be:

1. Discriminating ringing, where internal and external calls had a different ring cadence to alert the called party to the type of call waiting to be answered.
2. Intrusion on a call already set up, unless any of the attempted intruding parties had secure statue to block intrusion.
3. Manager-secretary facilities, which enabled call filtering by the secretary, but with manager able to contact the secretary

and automatic switch through to the manager when secretary is absent. Voice calling by secretary or manager and possibly auto answer by either if handsfree telephones are in use. One key calling of either manager or secretary. Visual indication of either manager's or secretary's line by either one.

4. Private line working, where a special line can be directed to a senior manager for undisturbed confidential calls.
5. Account coding of calls, either forced by the user on initiation of a call (useful in situations where consultants/partners must charge all clients for calls) or account coding at will, where the user can enter an account code during either an incoming or outgoing call to charge the client.
6. Broker's call, where the telephone user can hold two calls and shuttle between each call with secrecy on each line.
7. Callback on busy and call back on no answer, are all useful everyday features for the general telephone user.
8. Call pick up, which enables a user to pick up a colleague's ringing telephone from his/her own telephone.
9. Diversion of calls at various stages is helpful for efficient office organisation, as most can be preprogrammed so as to divert on busy i.e. when the called telephone is busy calls are diverted to an alternative telephone. Divert on no answer enables the user to divert calls on no answer to perhaps an alternative telephone than the divert on busy one. Divert all calls ensures that when a telephone is completely unattended e.g. out of office hours, all incoming calls are diverted to either another telephone, a voice mail system or a message centre.
10. Multiple party conference is a desirable feature to enable more than two parties to be engaged in a conversation. The ability to dynamically alter the number of parties engaged in the conference during the call makes the feature much more useful in the business environment.

4.6.2 Outgoing call handling

Outgoing call handling used to be relatively simple on a PABX; the software was required to check if the originator of the call was permitted to dial that specific type of number e.g. national, international etc. and then place the call with a free trunk handling software module and permit out dialing to commence. If the PABX had central speed dialling stores then these could also be handled by the outgoing call handling software and frequently the user would be permitted to speed dial international or national numbers but not make any other calls of this category.

Recent years has seen dramatic changes in the way outgoing call handling software has had to be structured. The software now needs to decide which carrier to use e.g. in the UK users have the choice of several competitive carriers or their own private network, with breakout at the nearest point to the destination on the PSTN.

The number dialled by the PABX user can bear absolutely no relationship to the addressing sent by the outgoing call handling software.

For example if the call is via Mercury for all national calls then the outgoing call handling software is required to check the look up tables built into the PABX to determine if Mercury service is available and appropriate, then to send the Mercury access code to the trunk handler, then send the Mercury PIN, and then the routing digits.

The software of the outgoing call handler is made even more complicated if an automatic route selection (ARS) or least cost routing (LCR) suite of software is included in the PABX, as this software contains numerous algorithms for routing calls based on the time of day, the load on the private network, and the charges being applied to the calls by the main carriers at that particular time.

An outgoing call request has to weave its way through all this software before the digits can be dialled out from the PABX to place the call.

The outgoing call handling has a further subdivision to cater for private networking and access the data base with the private network numbering schemes and routing plans.

For example banks use easily remembered numbers to access other banks and as all banks have a unique sort code then the PABX must be capable of translating between the sort code dialled and the routing information via the private network.

4.6.3 Incoming call handling

Incoming call handling software caters for a number of features just like the outgoing call handling. In the traditional PABX an incoming call signal was presented to the incoming call handling software and the software placed the call in a queue waiting for a telephone operator to answer the call before presenting it to the telephone or terminal handling software. Although this basic function still remains the software is required to handle much more complex tasks. The call may be a DDI call and therefore routed directly to the users telephone, or this may be a group of telephones, such as a service desk, where a number of users may be programmed to accept the calls. Only if the call is not answered will it be routed back to the telephone operator. The system may have a voice response attached to the PABX or as an integral feature of the PABX. In any case the call will first be placed with the voice response unit to answer, and provide the caller with the initial greeting, before either placing the call with the operator for onward processing or accepting DTMF signals from the caller to route the call to a particular port on the system.

The PABX may be a multi-tenant system and the operator may handle the calls for more than one company, in which case the incoming call handling software will be required to place the call in the appropriate company's queue for the operator to answer and flag the operator with the company name at the time the call is presented.

In many business applications the need for automatic call distribution (ACD) working is essential (Goodeve-Docker, 1995; Communicate, 1995). This facility originally required specially designed high traffic non-blocking systems with specialised control to route calls to agents who handled calls depending up the particular distribution selected by the business management. The ACD system could comprise a high traffic PABX front ended by a call sequencer and a management system that displayed the calls waiting in the queues on special screens. As PABX are now nearly all digital and non-blocking the PABX incoming call handling software has evolved to provide the functions of the ACD and call sequencing system as part of the incoming call handling task. The software is required to accept the incoming call request, route the call to a voice announcement, keep track of the time the caller is listening to announcement and switch the call onto a more appropriate announcement depending upon the time spent in the queue or the status of the agents.

The software is also required to interact with the telephone handling software to determine when agents become free to handle the next call and then force feed the call onto the agent's headset without the need for the application of ringing tone. The incoming call handling software is also required to interface with the real time call statistics software, which is a feature of ACD operation. The status of all calls in the queue, the number of calls lost, how quickly the calls are being answered, the mean time of individual agents to deal with customers, are presented as management information for the ACD supervisor to adjust the staffing levels of the system and therefore dynamically change the incoming call handling parameters of the system.

4.6.4 System reports and logging software

The software described so far has all had to operate in real time, but there are a few areas within the PABX where the operation is not time critical and therefore the tasks can be allocated a lower priority. These tasks are mainly concerned with logging of calls, generating reports on the status of the system etc.

An important part of any PABX is the measurement of the traffic that it handles and the call logging software does this. Generally call logging software will monitor the start of any call, who originated the call, where the call was destined i.e. the dialled digits, the duration of the call and the date and time of either the start or finish of the call.

The PABX will normally only provide a cyclical buffer memory of a few calls, as it is normally anticipated that the user will attach a printer or, more usefully, some form of external storage medium to record all the calls made in a given time period and then analyse them off line on a PC. The same logging software is also used for incoming calls but unless the incoming line is DASS2 with calling number identification then the log is restricted to the called extension. This can give time to answer by the extension, as well as the follow on call.

That is, the extension transfers the call to another extension plus the call duration and time and date of the call.

System reports are varied but the software will be expected to generate an equipment map showing the location of all the circuits within the system and whether the circuit is in service or has been taken out of service.

The software will also be expected to log all the problems encountered and generate trouble reports e.g. that a particular circuit has been seized numerous times but released immediately without call completion, indicating a faulty circuit which may be, for instance, the private tie line. This indication can be used to pass messages to the outgoing call handling software to remove the circuit from service and raise an alarm.

4.7 Data transmission

The PABX is not the ideal mechanism for data transmission as it is designed around the speech bandwidth of 4kHz and in digital switching uses circuit switched digital signals at 64kbit/s, which is not the ideal medium for LANs running at 10Mbit/s or FDDI at 100Mbit/s. However, for the occasional data user, where small amounts of data are transferred from one point to the other, it can make an economical medium.

PABXs have therefore concentrated in this area of the data switching market, providing such facilities such as modem pools on the PABX to which users can have dial up access, and calls set up over the PSTN with the modem.

In PABXs which have digital transmission two types of data facility can be provided. 64kbit/s switched connections, where the data circuit is normally of a slower speed and is rate adapted by a terminal adaptor using either an ECMA or proprietary rate adaption technique, which converts the data terminal output into the 64kbit/s bit stream. D channel packet switching is available on some ISDN PABXs, particularly French designs, where low speed data in packet format is supported on all basic rate ISDN lines and makes a very simple and cheap way to access packet network services.

4.8 Cordless PABX and key systems

A cordless PABX (or key system) may be simplistically regarded as a PABX (or key system) where some or all of the terminals are connected to the central switching system by cordless means instead of by hard wiring (Scales, 1993; Van Dussen, 1995; Bartley, 1995). Such a definition encompasses the situation where individual wired terminals connected to a conventional PABX are replaced by individual cordless telephones, and thus requires further qualification.

The additional requirement is that there should be some integration or interaction between the normal PABX functions and the cordless functions, in order to provide additional benefits or advantages. Such advantages might include smaller PABX switch requirements due to concentration within the cordless sub-system, or terminal mobility.

The cordless link may be provided by ultrasonic, infrared, or radio techniques using either analogue or digital modulation methods, but in order to meet the wide range of user requirements, digital radio techniques are almost invariably employed. The main advantages of cordless systems over their wired counterparts are:

1. Mobility, where a user is able to make or receive calls anywhere within the overall coverage area.
2. Reduced cost of ownership, due to simpler installation and ease of moving system users.

4.8.1 Requirements

The main requirements for the cordless sub-system to support the essential cordless PABX features are:

1. Support for high user densities.
2. Good subjective speech performance.
3. High reliability and grade of service.
4. Data capability.
5. High security.
6. Support for user mobility.
7. Self adjusting frequency allocation and re-use.

In environments such as city business centres with adjacent multi-storey buildings, fully equipped with cordless business systems, traffic densities of the order of 10000 Erlang per square kilometre per floor are required. Good subjective speech quality comparable to that of wired terminals is required at the cordless terminals, and the performance of the trunk network is often used as the yardstick.

Cordless business systems must be designed so as to provide acceptable grade of service (GoS) with respect to both call set up and forced call curtailment. Call set up GoS is a blocking phenomena as in conventional switches, whereas forced call curtailment GOS is a phenomena of cordless systems only due to interference of adjacent apparatus, and is subjectively the more important of the two.

Most modern PABXs and key systems are digital in nature and therefore inherently capable of providing both voice and data services. Cordless business systems must be capable of supporting the same range of voice and non-voice services. High integrity authentication procedures and mechanisms are required in order to provide secure access to the cordless system and provide confidentiality of the information transmitted over the cordless link.

A major feature of cordless systems is the provision of mobility both within a system and between systems, and specific mechanisms and protocols must be defined to support this important feature.

Wide area coverage systems such as cellular radio require central control and allocation of the channels available to each transmitter, to optimise channel re-use in order to maximise spectrum usage. This central control is not possible in the cordless business system area where a multiplicity of unrelated systems installed by numerous service providers must coexist. Cordless business systems must use dynamic channel allocation techniques to be self adjusting to changing local conditions.

4.8.2 Implementation

Although cordless PABXs and key systems are in their infancy, current designs are invariably based on radio links using digital modulation. The permitted radio spectrum bandwidth is shared amongst a number of users, with multiple access by the use of frequency division and/or time division techniques. In order to help achieve the high user densities required, a picocellular base station arrangement is used with cell radii ranging from about 10 up to 200 metres. Since all channels are available to all base stations by the use of dynamic channel allocation techniques, the traffic handling capacity of the system can be increased simply by adding extra base stations.

The radio performance in such systems is limited by the interference levels rather than by noise performance, so as the base station density increases, the interference levels from surrounding cells increase, and the usable cell size, which is determined by the carrier to interference ratio, automatically reduces.

Another possible method for sharing spectrum among a multiplicity of users is by the use of code division (spread spectrum) techniques. In CDMA systems the available bandwidth is not divided into separate channels. Each user has a unique code known to both handset and base station, which is used to spread the voice signal over the whole available bandwidth at the transmitter and to reassemble the transmitted signal at the receiving end. Although such cordless systems are complex compared with analogue systems, the low powers used, and the extensive use of digital techniques lead to highly integrated products. A block diagram of a typical CT2 handset is shown in Figure 4.12.

4.8.3 Standards

Commercial pressures towards open standards lead to the adoption of common air interface standards to encourage competition and facilitate roaming and public access services. The first two common air interface standards for cordless systems are CT2 and DECT and are both European in origin.

CT2 (second generation cordless telephone) and DECT (Digital European Cordless Telephony) both provide cordless speech bearers by using 32kbit/s ADPCM (adaptive differential pulse code modulation) to the ITU-T G.721 standard developed for transmission on the trunk telephone network. The ADPCM coding takes the 64kbit/s

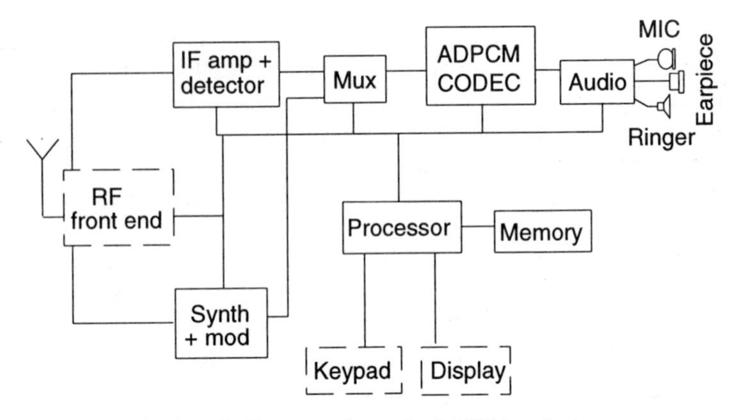

Figure 4.12 Block diagram of a typical CT2 handset

PCM bit stream comprising 8-bit samples at the 8kHz rate, and reduces the 8-bit samples to 4 bits.

CT2 uses a 4.0MHz block of radio spectrum between 864MHz and 868MHz to provide 40 duplex channels, and DECT uses the 20MHz from 1880MHz to 1900MHz to provide 120 duplex channels. Average power levels are around 10 milliwatts per channel for both systems, and both systems employ dynamic channel allocation.

There are substantial differences in the frame structures of the two systems as shown in Figures 4.13 and 4.14. CT2 is a frequency division multiple access, time division duplex (FDMA/TDD) system arranged as 40 frequencies spaced at 100kHz intervals, with one duplex channel per frequency, using time division to carry the traffic in both directions. In order to interleave the transmit and receive bursts and maintain the overall bearer capacity, the transmitted bit rate must be at least twice the overall data rate. For CT2 the transmitted bit rate is 72kbit/s.

DECT is a multiple carrier, time division multiple access (MC/TDMA) system which uses ten channels at 1.728MHz intervals, with 12 duplex channels multiplexed on to each carrier and a transmitted bit rate of 1152kbit/s. Within a DECT base station each time slot may be transmitted on any of the ten carrier frequencies, and a single base station may operate with up to 12 handsets simultaneous-

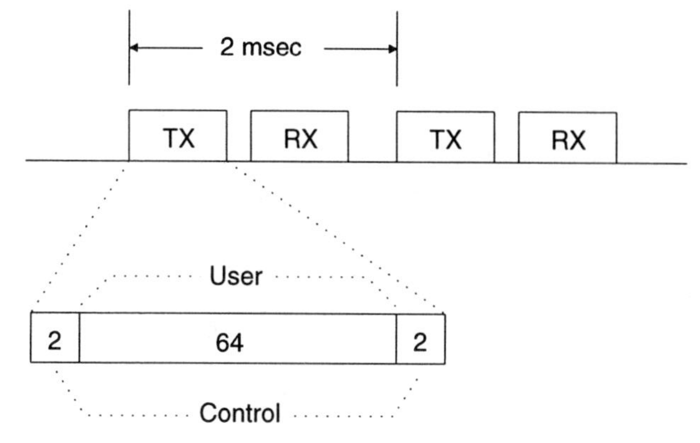

Figure 4.13 CT2 frams structure

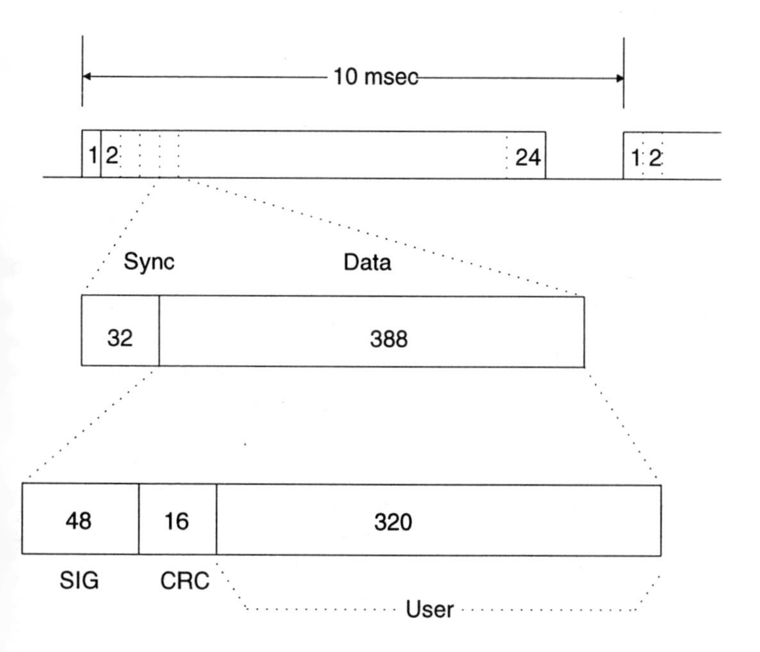

Figure 4.14 DECT frame structure

ly. In both systems a transmit or receive burst comprises a number of ADPCM samples together with control and signalling data.

The use of time division techniques introduces delay in the speech paths which may require additional echo control to provide acceptable subjective performance, particularly in DECT systems with their longer framing delay.

Both CT2 and DECT specifications have adopted a layered approach based upon Layers 1 to 3 of the ISO open systems interconnect 7 layer model. These cover the physical aspects such as carrier frequencies, power levels, modulation characteristics, multiplexing of user and signalling information, channel assignment, link set up and control, and network layer protocols for managing calls between the portables and the central system and external networks. In addition the requirements for speech transmission performance over the air interface are also specified.

4.8.4 Possible configurations

Figure 4.15 shows a simple cordless PABX structure where N standard PABX extensions are assigned to N cordless handsets. The radio

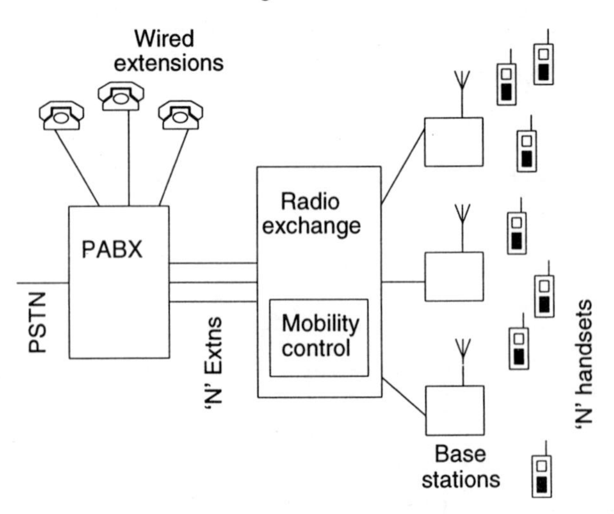

Figure 4.15 A basic cordless PABX arrangement

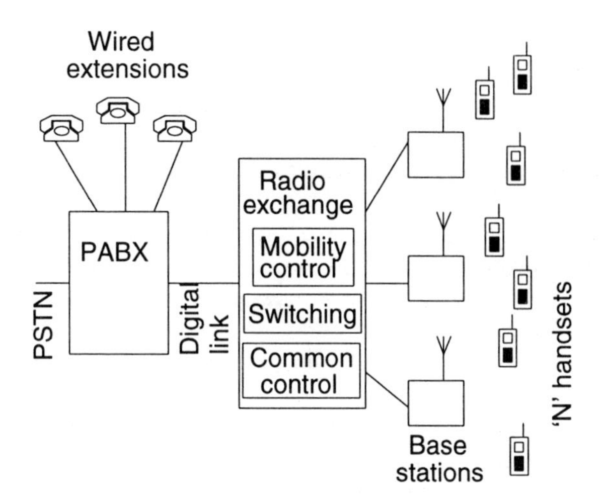

Figure 4.16 An improved cordless PABX arrangement

exchange links the PABX extensions to the radio base stations and also handles handover and location updating functions. The PABX provides the normal services. This configuration may be added to most existing types of PABX.

Figure 4.16 shows a more complex structure where the radio exchange incorporates additional common control and switching functions to improve roaming and to provide concentration. The connection between the radio exchange and the PABX may be by a multi-channel digital link, such as ISDN primary rate access. The normal services are still provided by the PABX. The radio exchange and the PABX may be built as a single integrated system or as separate unit.

4.9 References

Armstrong, L. (1995) End of the line for PBX? *Communications International*, February.

Bartley, G. (1995) DECT awakens the sleeping giant, *Mobile and Cellular*, February.

Communicate (1995) The key to better customer relations and revenue, *Communicate*, May.

Goodeve-Docker, J. (1995) The integrated call centre of the future, *Voice International*, April/May.

Postlethwaite, D. (1990) Introducing the WPBX, *Communications*, November.

Prentice, S. (1994) The PBX as a Universal Access Platform, *Telecommunications*. March.

Scales, I. (1993) Next stop: the wireless home & office, *Communications International*, June.

Thomas, M.W. (1991) Trends emerge in PABX technology, *Communications International*, July.

Van Dussen, B. (1995) Cordless PBXs give room to roam, *Communicate*, February.

5. Centrex

5.1 Introduction

Centrex is a service offered by telephone companies which gives their customers the capabilities of a sophisticated voice and data Private Branch Exchange (PBX) without the need to purchase and maintain on-premises switching equipment.

A modern digital PBX is a specialised real-time computer which routes calls and provides calling features such as call forwarding and handles data transmission for a customer's business site (see Chapter 4). Lines connect each individual telephone and data terminal to the PBX. The PBX is connected to the Public Telephone Network Operator's (PTO) main exchange (called the Central Office switch) by voice and data channels called trunks. Voice and data traffic to and from the public network travels over these trunks, while traffic internal to a customer's site is routed by the PBX. Traditionally, trunks do not provide any features.

Centrex is an alternative for businesses which do not wish to buy and run their own private telecommunications system (Sazegari, 1993; Jones,1994). As the name implies it is a Central Office Exchange, with the lines from individual telephones and data terminals connected directly to the PTO's exchange rather than a PBX. This exchange is also used for the public telephone service. The software resident in the central office switch is partitioned and programmed to create a virtual PBX that delivers voice and data services comparable to those on any modern digital PBX. Numerous virtual PBXs can exist on any one central office switch, all sharing the same processing resources.

Centrex is attractive to corporations, particularly the larger ones, who increasingly find that telecommunications is using up considerable resources, financial and human. This coincides with these organisations seeking to create strategic advantage in their core business activity using sophisticated telecommunications. In the UK this has resulted in a proliferation of private networks and large departments to manage the telecommunications service. An alternative service from the public network which fulfils the same requirements allows these organisations to outsource their telecommunications and concentrate on their core, revenue generating, activities. An additional benefit for multi-sited corporations is that their various sites can be linked together over a public telephone network based on sophisticated central office exchanges such as the Nortel DMS range, to form a 'virtual private network' (VPN). This gives a simpler internal dialling plan as well as network wide use of features such as call forwarding, ring again and call waiting.

Centrex is also attractive to business that do not have enough lines to justify the purchase of a PBX, yet need the sophisticated voice and data services an expensive PBX can offer, such as Automatic Call Distribution (ACD), and integrated Voice Mail (IVM).

5.2 Centrex services

The network and PBX approaches offer comparable features and benefits. When both options are available, as in North America, hybrid networks are frequently found, with a vast majority of Centrex users using a PBX somewhere in their network. However, PTOs can offer integrated access to a much wider range of service with Centrex, for example ACD (see Chapter 6), advanced data switching such as Frame Relay for LAN to LAN interconnection, and private leased lines. A typical deployment of Centrex and hybrid networking on the Nortel DMS is shown in Figure 5.1.

5.2.1 Featured voice

Centrex has voice features which help ensure that calls are answered quickly and routed appropriately and cost effectively within the network. Features provided by typical advanced digital Centrex include direct dialling inward (DDI), flexible intercept, hunting, night service and security provisions. Advanced business sets with user programmable feature keys enhance access to a range of features. The

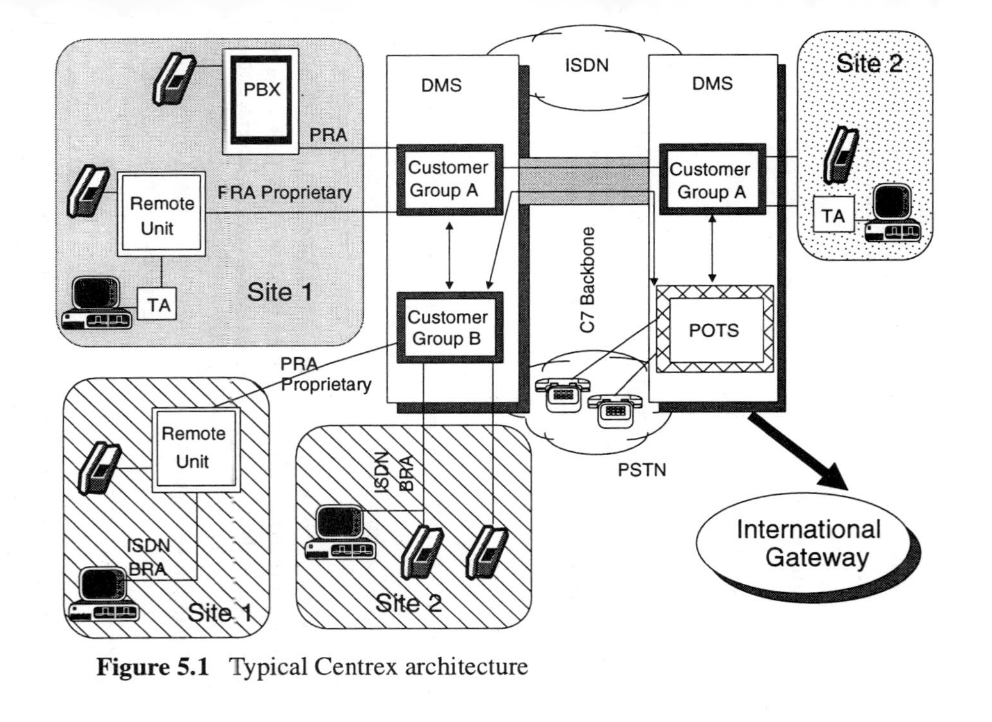

Figure 5.1 Typical Centrex architecture

addition of displays on these sets provides calling information such as the called or calling party's name and number, reasons for redirected calls, and digits dialled during call origination or while activating system features.

The system offers further facilities, including ACD with sophisticated management and centralised message service. A simplified console enables operators to concentrate on incoming calls.

5.2.1.1 *User terminals*

Nortel's terminals support three access interface technologies, namely:

1. Analogue, for plain old telephone service, or POTS.
2. Business set, for network-wide featured voice services.
3. ISDN, for network-wide interworking of voice and data capabilities.

These technologies complement each other to meet both the user's and the PTO's requirements for service and ease of deployment. The ISDN interface is an enhancement to the business set interface, which in turn offers an improvement in functionality over the POTS interface.

The attraction of the POTS interface stems from its historical standardisation and simple implementation at both the subscriber and central office ends of the loop. The limitations of the interface result from its single analogue channel. Consequently, multiple line appearances require a key system at the customer premises. Limitations of the interface also make access to most Centrex features difficult; the user has to use 'feature' codes, for example, '*74' is used to invoke call forward.

With the more advanced sets users can take advantage of assignable feature keys, which eliminate the need to remember the feature access codes. Some terminals also offer the user hands-free capability, message waiting features, line status monitoring and visual ringing. However, these features are dependent on what the terminal can support.

The POTS interface can transmit data only with a modem at limited data rates. It is, however, the most cost-effective interface for basic telephone service.

The second access technology, the business set interface, offers significant advantages over the POTS interface. The benefits result from an out of band signalling channel which carries the feature control signalling information independently of the voice call over the same pair of wires.

This capability to carry both voice and signalling information allows multiple line appearance on the terminal and provides users with easier access to advanced Centrex voice features. The out-of-band signalling capability also allows the switch to constantly update the terminal on the status of the line or any feature being invoked.

This technology was first introduced by Nortel in the early 1980s. One of the most recent additions is the M5000 series of advanced business sets, offering single key access to DMS digital Centrex features such as ring again and three-way calling. User response from US and Canadian field trials has been extremely positive, particularly in the areas of audio quality, functionality of the display and ease of use.

Most of the functionality of these sets resides in the DMS-100 software. The sets themselves transmit messages corresponding to key depressions to the switch, and receive indicator messages from the switch. Because the terminals are unaware of the feature actually invoked, new features can be added and old ones upgraded without changing the installed base of terminal hardware or firmware.

The introduction of liquid crystal displays (LCDs) on some of the business sets further improves the user interface by enabling, where available, the caller's number and name to be displayed. There are also other reasons for the display, such as allowing the user to see why a call has been forwarded, and from whom.

It is frequently postulated that most people in a business environment only use about four features on their telephone system, mainly because access to these features is inconvenient. The M5000 series makes it much simpler to use many features available on the DMS Centrex, thus allowing normal users to realise the benefits these features have to offer.

5.2.2 Automatic call distribution

Automatic Call Distribution (ACD) is a powerful business feature which can be readily implemented in Centrex. It distributes large numbers of incoming calls equally among a designated group of answering positions, presenting the first incoming call to the position which has been idle longest.

If all 'agents' i.e. answering positions, are busy, later calls are queued and answered in order of arrival. Comprehensive information on the incoming calls, such as time to answer, number of callers who abandoned while in the queue are all available from a management information system.

Current DMS systems will deliver a full ACD feature set regardless of line size, whether to a 2000-line sales or reservation system or a 10-line group medical practice. This extends the benefits of ACD to small and medium sized businesses which would find the cost of a PBX-based solution prohibitive.

The DMS also allows network wide distribution of ACD groups. A service organisation e.g. motor insurance, could publish one directory number for inquiries, but incoming calls would be passed to an agent group located closest to the caller, reducing the cost of the call. It also allows calls to overflow to other groups if the first group is busy.

Using the calling line information passed with each call, sophisticated applications can be developed e.g. for telemarketing or information services. This is done via a Switch to Computer Applications Interface (SCAI), which allows the DMS to be linked to business computers.

These services can be used to generate extra revenue (e.g. mail-order) or save on operating expenditure (e.g. outbound calling to check customers are at home before sending out service engineers). Further sophistication can be added by using an Interactive Voice Response system, with the ACD, to provide interactive routing, such as determining if the caller wants sales or service departments before queuing the call. Credit card validation systems also use the same technology, which can, for example, be provided on the Nortel Meridian Mail system.

5.2.3 High speed digital data

Convergence of voice and data in digital networks is an accomplished fact. A recent US survey of Fortune 1500 companies showed that despite the continued dominance of voice traffic in terms of volume, data is the dynamic service requirement and driving force behind new applications.

Local Area Networks (LANs) solved immediate data handling requirements at working group level. Local integration of voice and data at site level through a PBX delivered greater interconnection options. The use of private leased lines to interconnect sites created wide area data networks.

Digital Centrex brings the flexibility, capacity and resilience of the public network to wide area data networks as well as Integrated Service Digital Network (ISDN) capabilities.

Major offerings available from data services implemented on Nortel DMS digital public switches are:

1. Datapath services provide completely digital synchronous and asynchronous data transmission with automatic rate adaptation and handshaking protocols, freeing the user from the need to get transmission parameters for each call. Datapath will also work into analogue data transmission systems through a DMS-based modem pool
2. DataSpan, an integrated Frame Relay, based on standards recommended by the ITU-T. Frame Relay is a wide area 'fast' packet switching technology that offers an intelligent alternative to private lines for data interconnection. It uses less error checking than packet switching, optimising on the high quality of transmission on digital networks. This makes it more efficient for applications such as LAN interconnection. Existing solutions for connecting several data sites require leased or dedicated lines of fixed bandwidth in a complicated mesh. With DataSpan the public network becomes a viable alternative, eliminating the need for multiple connections; each location now links to the other through a single access line to the Frame Relay network.

3. ISDN services which realise the full potential of data management by Centrex. Standard features are being implemented on the DMS as they are defined. The Basic Rate Interface (BRI) carries two 64kbit/s B-channels plus one 16kbit/s signalling channel and delivers voice, circuit and packet switched data to as many as eight terminals on one pair of telephone wires. The Primary Rate Interface (PRI) offers 30 B-channels plus one 64kbit/s signalling channel, and can form the ISDN link between a central office switch, digital PBXs, and host computers.

Since all DMS data services provide switched access, terminals can reach multiple hosts as quickly and easily as dialling a normal phone call. Digital Centrex provides a range of calling features as well as a range of security features, such as defined customer groups, closed user groups, and direct inward system access authorisation codes, to guard against accidental or deliberate breaches of security.

The DMS also has the capability to support existing simple data networks based on drop and insert multiplexors utilising spare capacity on private leased lines for voice networking. Changing the voice network over from a PBX to Centrex will not call for an immediate re-appraisal of the overlay data network.

A key benefit resulting from taking the Centrex route to ISDN is future-proofing of the corporate network. New ISDN features are continually being defined and are likely to force expensive upgrades on PBXs. In contrast, upgrades by the PTO to central office switches are immediately available to all users.

5.2.4 Network management

The existence of a virtual private network means that the telephone company is able to offer customers advanced network management features. The user specifies the features required and retains direct control over their use. In addition, the operator can provide central monitoring, measurement and control, allowing more sophisticated management than possible on a network of PBXs. The management systems will capture comprehensive details of every call, which can be sent to the customer with the regular bill or transmitted on a data link directly to the customer.

5.3 Development of the Centrex market

By the 1960s AT&T, the North American PTO, had achieved universal POTs service. They then looked for new market opportunities and Centrex was born.

First generation analogue Centrex services were highly acceptable by the standards of the day, and, by the start of the 1970s, 20% of the business market used Centrex. This then slumped as AT&T chose to focus on the PBX business to protect its position in anticipation of divestiture. Centrex recovered sharply following the divestiture of AT&T and the introduction of more highly featured digital Centrex such as that on the Nortel DMS. Today, every Bell Operating Company has experienced growth in Centrex services, which once again handles close on 20% of North American business telecommunications.

In Europe Centrex is emerging as a result of liberalisation, with the first European digital Centrex service launched in the UK by Mercury Communications in 1987. Mercury regards Centrex as a technique for adding value to its public network services, while simultaneously delivering competitive advantage to customers.

The market's initial perception of Centrex as expensive led to some customer resistance. However, in 1989 Peat Marwick, the accounting and management consultants, compared the total costs of Centrex with PBXs and concluded that in four UK company models calling for between 100 and 1500 extension lines, over seven years Centrex turned out to be the economic choice, with the greatest savings made through operating efficiency.

Mercury's initial service was limited to customers directly connected to their central London optical fibre network. In four years, features have been enhanced to make sure that the service continues to meet or exceed leading PBX offerings and tariff reductions have made the service even more competitive against PBXs. Early users included members of the London financial community, who saw customer benefits as:

1. Speed of installation and service; less than 4 months from signing of lease to full operation in a new building.
2. Financial flexibility; the equivalent of a £2M installation without capital outlay.
3. Operational flexibility; the ability to add or alter functions, and to cope with expansion and possible office moves.
4. Maintenance; 24 hour monitoring and support.

The next logical step is wide area Centrex. This makes it possible for the user to configure a highly featured wide area virtual private network while retaining an exceptional level of flexibility. Wide area Centrex draws on a database of users, identifying their location, and privileges, giving them the full range of Centrex business services at all sites. It amounts to a virtual private managed network operated by the telephone company retaining the availability possible only on the PSTN.

The telecommunications revolution has so far served the business user within the office environment. Extension of the wide area Centrex concept through intelligent networks brings the networked office environment into the home and also brings the home worker into the mainstream business community through shared facilities. Any user, wherever he or she is, can have access to the range of services which would previously have been available only in the office environment on the corporate network.

The advantages of Centrex for the business and domestic consumer are clear, such as variety of features, low investment, protection against obsolescence, reliability and flexibility to meet change positively. In addition, it greatly strengthens the position of the telephone company as a one-stop service provider. The basic telephone network is unique in its ubiquity and vast resources. Centrex enables numerous services to be added, or deleted according to business volume and type of activity.

Given this range of service flexibility, the traditional understanding of the term Centrex is limiting. The DMS allows telephone companies to deliver Network Based Services thereby positioning themselves as a responsive business supplier rather than a commodity service provider.

5.4 References

Jones, D. (1994) Moving on to Centrex, *Telecommunications*, May.
Sazegari, S. (1993) Digital technology resurrects Centrex, *TE&M*, 1 May.

6. Call management

6.1 Introduction

Call management began as the recording of basic call information by sampling signalling activity on voice trunk circuits carrying calls into and from a call centre.

The information provided circuit usage, but not an identification of who placed the call and to where it was placed. The information may have been derived by equipment at the facility or from the public phone office providing service to the facility. Analog printers, magnetic tape recording devices, or mini computers were used to perform the recording.

The carrier may have collected and charged for the information on a request basis, with the customer responsible for processing the data. Basic reports of not much more than call by call activity were created from this information. Visual scanning of the reports for exceptions and other management information was then performed.

Call management has continued to grow in many dimensions from its humble beginnings (Huffadine, 1990; Communicate, 1993). Competition among switch vendors, the advent of the PC, plus demand from call managers for more information, more quickly, are all contributing to new systems and management capabilities.

This chapter covers voice call management in its current dimensions and looks at trends for the future.

6.2 Why call management?

The desired goal of effective call management is almost always to reduce costs and to increase the profit for the call centre (Costello, 1990; Communicate, 1995; Hayday, 1995).The switch hardware, the carrier services and connections, plus the user calling habits are the major areas which affect the costs and profit.

The ways in which the goals are achieved can depend on the type of call centre and the type of call management used. The call management approach, with the investments in systems and services to be used, should be carefully considered against the types of call centre and the management goals in the short and long term.

The payback for investments in call management systems, software, and personnel can be cost justified on one or more elements to be optimised. The method for optimisation to produce the greatest profit is dependent on the application.

In an application providing help for product service related calls, the goal would be to keep all calls as brief as possible so time sensitive trunk circuits are used less. In another application which provides consultation and advice, and bills callers based on time, the goal could be to encourage continuing the conversations, since the longer the call the more the profit.

In other applications the goals may be to reduce the number of calls or to allow more calls to be made. The reduction of excessive calls, elimination of under utilised extensions and trunk circuits, plus improvements in service are other goals useful as cost justification elements.

A common approach used in cost justification of an ACD (Automatic Call Distributor) to be used for high volume inbound call applications is to assign a cost or profit value to each call, or even to each second in time of call duration.

The investment in the completely stand alone ACD can be proven just by using the improvement in answering time, compared to other alternatives (Farwell, 1991; Johnstone, 1992; Gray, 1992; Sinden, 1994). For example, if the application is moved from a PABX group to the ACD, the ACD can answer each call and connect it to a service person faster than the PABX.

The expected payback period is the result of dividing the cost of the ACD by the product of the difference in seconds of answering time, the number of calls per second and the value per second. If the payback period is short enough, then the cost is justified for the call management benefits the ACD will provide.

6.3 Special needs

Under the general call management sphere there are individual environments which are generally defined by the type of call application. Each environment has special call management requirements. In high call traffic inbound and outbound telemarketing applications, management of the trunk circuits and call servers is as important as handling the calls.

The stand alone ACD system, and to a lesser extent, the ACD option packages in PABX systems, are specifically designed for these high traffic applications. In other applications, multiple environments are being addressed by common hardware and software programs which are tailored to the application by adding options. Several of the major environments will be considered in detail.

6.3.1 Corporate environment

In the corporate environment (refer to Figure 6.1 for a general representation of this environment) call management is directed at several distinct areas of interest. The most prominent area consists of PABXs providing internal communications and connections to non-corporate systems. The call traffic is basically intracompany and intercompany business calls, plus the non-business related calls. The next areas are those groups dependent on call traffic, providing service and generating revenue to the corporation. These areas are PABX attendants, ACD groups within the PABXs, and stand alone ACD systems. The final area appears when the corporation occupies multiple distant locations. This area connects the PABXs and ACDs at the multi-site locations by a company private network of tandem switches or by virtual private network interfaces. Each area in this corporate environment can be addressed by separate call management systems, reports and management approaches.

Call management in the corporate environment PABX area should first be directed at controlling calls before they are made. This is accomplished by providing employees with phone books covering good calling practices and listing restricted numbers, letting them police themselves, and then by using internal call restriction tables in

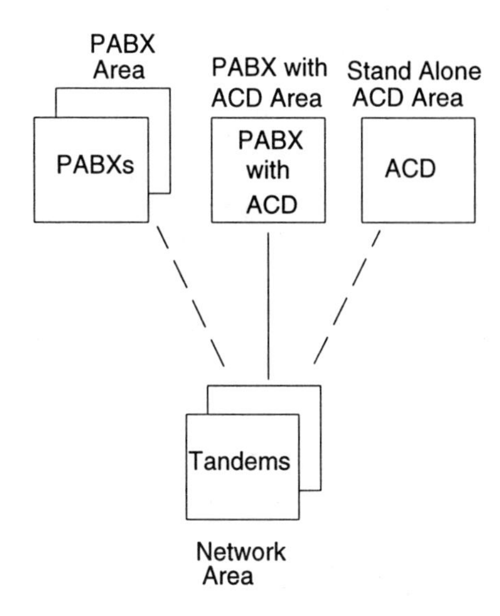

Figure 6.1 Corporate environment with areas of call management

the PABX to catch or block all violations. The internal call restriction tables would contain specific and partially specified phone numbers. For example, one partially specified restriction may be for all numbers in an entire country code, another for a selected city within a country, and another a specific number within a city. The restrictions, depending on the PABX, are specified as either the numbers allowed or not allowed to be dialled. If the PABX can assign a class level to the users, then the restrictions could be applied to all, or selected class levels. Such restrictions are sometimes based on date and time. The PABX would use the day of week and time of day to map to a specific table to be used from a set of restriction tables. The restrictions might also be based on the trunk circuit type for which usage or special service charges would be incurred.

Information on the number of attempts to restricted numbers can provide valuable information. Network modifications and removal of restrictions may be required if legitimate attempts are being blocked, but are being completed by other methods at a higher cost. The call restriction tables should be easily changeable and should be regularly reviewed to reflect current requirements and call management goals.

Calls which pass the restriction tables should be analysed for traffic patterns, call durations, and other parameters useful to call management. This information is derived from the call detail records produced by the CDR (Call Detail Recording) or SMDR (Station Message Detail Recording) features of the PABX. The PABX either produces the information in report form directly, provides the detail to a storage media for off-line processing, or outputs the detail to an external reporting system through a V.24 interface. All of these alternatives and the reports produced from the call detail are covered in more depth later in this chapter.

While the PABX typically provides detailed information for outbound calls, it can be very useful to capture detailed information for inbound calls also. This is sometimes available as a special service, usually called AMA (Automatic Message Accounting), from the carrier providing calls to the PABX. This information can indicate a need to add more trunk circuits in some areas to reduce blocking, and to remove trunk circuits in other areas because of low usage. It can also be used to find abuses, such as excessive calls to selected extensions.

Call management in the corporate environment network area is directed at reducing the costs for the network facilities and the costs per call. Privately owned network facilities, leased, public and common carrier services may all be used at times, with changes made to obtain the lowest overall cost based on the tariffs for the call volumes offered to the network. The other area for call management attention is to reduce employee abuse and eliminate non-employee usage of the network.

Calls are offered to the network area from directly connected PABXs and ACDs, plus from public and common carrier networks. Direct access to the network may also be provided from carrier networks so that employees on travel or at home may perform business transactions. Such access must be protected from misuse by unauthorised users and non-employees. In the most common forms of protection, the network access numbers are disclosed only to authorised users, and individual employee identification numbers are assigned which must be dialled with each call. The access numbers are changed periodically as a further preventative step.

The tandem switches in the network should validate identification numbers before access is allowed, then provide automatic least cost routing for approved calls according to the caller's privileges on the network. The alternate route choices are ordered and automatically tried by increasing cost for the time of day and day of week. The caller identification, the route choice, plus other call information is entered in the call detail record. From the call detail and circuit utilisation information produced by the tandem switch the call management decisions must be made.

In the corporate environment, high volume service and revenue producing call traffic is handled most efficiently by ACD functions. The ACD functions are provided by optional add-on features in the PABXs or by stand alone ACDs. Call management in the ACD is directed at the equipment, trunk circuits, call handling staff, application groups, and software controls on call handling. The management decisions are made from real time display information and from detailed reports produced directly by the ACD. Often the decisions must be made and actions taken very quickly to react to quickly changing traffic patterns. Longer term decisions can be made from the daily, weekly, monthly, yearly and call detail reports produced by the ACD.

Basic ACD design is founded on the management of queues. Incoming calls enter a queue and wait if servers are not available. If servers are available but there are no incoming calls, then the servers wait in queue. A very typical goal is to connect 85% of all incoming calls to servers within 20 seconds. Real time display information and reports produced every 15 - 60 minutes show management how close to this goal actual performance is.

If performance is better than 85%, then staff may be re-assigned to other functions. If performance is less than the goal, then additional staff may be required.

Call management in the ACD is often divided between staff managers and a network manager. The staff managers have to use the alternatives available to adjust the staff levels to meet the performance goals under the current traffic load. Reassigning staff between applications, from less busy to more busy, may be the first alternative. This should not require hardware changes, but it must be possible by software commands to the configuration database. A resource pool of part time employees who are willing to work flexible hours as needed, is another alternative. Another choice may be to let an interactive call response unit record caller information with the intent of the servers calling back as soon as the current traffic peak passes.

Managers in applications using more than one ACD may have another alternative to balance traffic. The ACDs may be able to automatically transfer a portion of the calls waiting for service to each other to balance the traffic loads. This is especially effective when the ACDs are in different time zones with peak traffic somewhat time dependent.

In this manner the ACDs work to provide better individual performance and better performance over all the network. The network manager determines the most effective ways to move the calls between the ACDs.

If staff adjustments or other alternatives cannot be made, then management may attempt to limit calls coming into the system. A queue limit or waiting time for service can be established. When the limit is exceeded an announcement is played to the caller indicating the expected wait time and alternatives they may have, such as calling back later, leaving a message, or transferring to another department. A more definitive action may also be taken in that if the established queue limit is reached, then any new calls beyond the limit receive a timed period of busy or re-order tone and are then disconnected. If callers immediately call back, then this alternative may actually be counter productive.

An external call management system may be connected to the ACD for staff scheduling and forecasting of staff and trunk circuit requirements processing. ACD applications commonly operate full time every day with full and part time staff. Scheduling and forecasting for this staff becomes a formidable task requiring the special system, especially as the ACD grows to the thousand or more positions supported by the current stand alone ACDs.

Call traffic patterns and staffing requirements for an ACD can be forecast based upon past history information with expected growth and management performance goals. The scheduling and forecasting management systems can provide information to know how many trunk circuits are needed, expected total call volumes, and what staffing levels are needed for future periods. A period can be even to the half hour detail. When connected directly to the ACD with a V.24 interface, actual performance information can be used to track and produce new forecasts.

The call detail information from the ACD can have special significance for both management decisions and for customer service billing. Information may be collected on both inbound and outbound calls. Answer supervision, either provided by signalling or by hardware detection of voice energy, is important to determine whether an outbound call is answered and call duration is correct.

6.3.2 Lodging environment

Call management in the lodging environment is directed primarily at appropriate sizing of trunk circuit groups and accurate answer detection for outbound calls. While inbound calls may follow a uniform distribution, outbound call attempts may often experience extreme peaks during the evening registration period and following the closing of large meetings.

General guest dissatisfaction can occur if it is consistently difficult to get an outside line and often necessary to go somewhere else to make calls. If guests are billed for outside calls, then it is important to know whether calls are only attempts or are completions. Making the determination based solely on call duration is an inaccurate approach, which opens the way for the guests to contest phone charges on individual calls. Answer detection performed with a high degree of accuracy by the PABX, or by an attached call management system monitoring the trunk circuits, becomes important to the hotel and guests.

With an interface to the front desk and direct connection to the PABX, excessive phone charges can be controlled. At registration an established credit limit for phone charges can be entered into the system. As calls are made, this limit is checked. When the limit is exceeded, the call accounting system informs the PABX to block further calls.

6.3.3 Hospital/University environment

The hospital and university environment also requires very accurate call billing information since users pay for individual toll calls. This environment and the corporate environment share the same call management requirements for least cost routing alternatives when large campus installations are created using multiple PABXs connected in a network.

Call restriction capability is less important, however, than in the corporate environment simply because the PABX is a revenue producer.

6.3.4 Client environment

The client environment is that segment of business where calls are to be billed to a client or account rather than to the person making the calls. The caller enters the account or billing code for the call along with the destination number. The PABX enters the code into the call detail information.

PABX attendants may also perform a similar operation if an outcall cannot be placed directly, but only by way of the attendant. Billing reports for this environment should provide separate reports organised by the code.

6.3.5 Shared tenant environment

In the shared tenant environment the call centre is a revenue producer, since users are billed for the calls they make in addition to basic services. Call management has to be directed at providing good service while keeping overhead costs low.

Circuit trunk group sizes, the optimum number of attendants, and optional services, such as voice mail, all have to be considered under shared and non-shared operation. In this environment the call centre and any call accounting systems that are used require shared tenant reporting capability.

6.4 Call accounting

Call management in the various application environments is heavily dependent on call accounting information (Costello, 1991). In order for the information to be useful, however, the large volume of information must be organised and summarised into manageable reports and displays. If the switching systems cannot perform this function directly, then there are many choices available for add on call accounting systems and outside services.

The add on call accounting systems and service bureaux compete for customers by offering a wide range of reports, formats, data selection methods, and display outputs. If the user wants a report which currently does not exist, then the system may provide online report creation capability, or the vendor may offer this as an optional service.

The standard list of report choices include: call by call detail by time, by extension and by trunk circuit, busy hour, plus summaries by extension, department, account code, trunk circuit, and trunk group. These same reports are offered in a weekly, bimonthly, monthly and yearly version. Bar, pie and line graphic charts, and histograms are common alternative formats to the standard analytical printed reports. Colour is often used with the graphic reports and displays. Rather than printing reports, the output may be written to magnetic media or retained on the system for manual scanning and selected report printing.

Call cost information may be included in the reports with the cost calculated from a current tariff database using V&H co-ordinates. The report may be used directly for billing or for comparison of used facilities versus the alternative tariffed facility.

Exception reports, if available, can provide very beneficial information. Exception reports can list those calls with call durations longer than a specified length or greater than a specified cost. Exceptions can also show those extensions where more calls than a specified amount were made, plus the total calls made to specified numbers. The reports can be used to detect misuse, high or low volume traffic patterns to areas which may require network reconfiguration and facility changes. Exception reports can also indicate switching hardware and trunk circuit problems. Trunk circuits which consistently carry many calls with very short durations, or very few calls but very long durations when compared to other circuits in the group, are indications of problems in the switch hardware or trunk circuit.

6.4.1 Service bureau

Service bureaux process call detail information into a finished product for return to the customer. The call detail information is typically provided to the service bureau on magnetic media, either tape or diskette, on a weekly or monthly basis. A more direct and timely interface may be available by daily polling from the service bureau to call buffer units at the user's switches to receive the information. The finished product is dependent upon the service bureau. Printed reports are almost standard. Magnetic media containing reports to be printed or modified by the user for additional information before printing is another alternative.

The advantages of service bureaux are that they provide the processing hardware, supporting staff, and can offer reports customisation services. Since the staff is dedicated and familiar with the product, new reports can be developed very quickly. The service bureau is also not affected by real time call traffic rates, since the data is processed off-line. Distribution of product to individual customer departments can occur directly from the service bureau.

The disadvantages of the service bureau are that flexibility and control may be lost. The customer is now dependent upon the service bureau's schedules, product revisions, and cost structure changes.

6.4.2 User owned systems

The alternative to the service bureau is for the user to own, or lease, the call accounting system. The hardware and software may be acquired from the same vendor; however, most PC-based software systems are sold independently. The user may also choose to perform the software development.

The advantages of the user owned system are that reporting alternatives are just as available as provided by the service bureau. The software vendor may offer alternative packages, or the package may allow report selection and creation online.

Reports can be produced more frequently, such as daily or on demand for selected call tracing. The user may also be able to add other software packages, such as front desk, attendant directory service, and PABX moves and changes to the same hardware.

The user must determine the type of hardware, the operating system, and the call management software to use, plus any other application software packages to be integrated. Assurance must be made that the operating system and application packages allow seamless integration so there are no conflicts which allow only one or the other to execute.

The user has to decide whether a PC-based system can handle the call volume and peak traffic periods, plus provide the reliability, or whether a mini or mainframe will be needed. From history or by estimates, a determination has to be made of how many call records will be created per reporting period(s) and the call rate during the peak traffic periods.

The user must decide whether the system will be directly connected to the call centre or whether it will interface through a polling unit. If polling is used, then the frequency of polling and the call rate will determine the requirements for the size of the call buffers.

There are other questions which have to be answered. Will the system be dedicated to one call centre or will it be connected to many call centres at remote locations? If there are many centres, then is there good cost justification for not using a service bureau? If online access will be allowed for report manipulation and perusal, then how many terminals should be provided? Will the terminals be directly connected, remotely located, or interface by LAN or WAN? Again, can a PC handle these requirements, or will a mini or mainframe be required?

All these requirements should be put in the procurement specifications. The vendor should be required to demonstrate performance against the specifications as one criteria for acceptance of the system.

6.5 System interfaces

Call management systems currently on the market almost universally expect to connect to the switch by a V.24 port. The port is an RS-232-C physical interface. The call management system can be co-located with the switch or remoted using modems. The older method of connecting to the trunk circuits may still be an available alternative to the V.24 port, but offers less information and flexibility to automatically handle trunk configuration changes.

If full time call management information is critical, then interface failure detection processes should be provided between the switch and call management system. Since producing call management information may be a minor objective for the switch, failure detection becomes the responsibility of the call management system. An alarm, either audible or visual can be generated if call information, or keep alive data, is not received as expected within a specified time period. If the detection is based on receiving call records, then the process should adjust to lack of records that may naturally occur during some times in the day.

The data buffer capacity in the switch and call management system should be considered, for this becomes an important factor during out of service periods. Standard buffer and optional expansion increments should be considered against the requirements. If buffer capacity is 10,000 records, the switch produces one call detail record per call, and the call rate is one call per second, then the buffer can hold call records for 2 hours, 46 minutes and 40 seconds. If the requirement is to support an out of service period of 4 hours, then a buffer capacity of 14,400 records is required.

Depending on the switch, call detail information is provided either as a single record containing all the information at the end of the call, or as a series of records which are produced as phases of the call occur. In the latter case the switch assigns an unique sequence number to each call. All the records pertaining to the call contain the sequence number. The call management system then must use the sequence number to locate and accumulate all the records for each call.

The call detail information may be provided in ASCII, binary, or a combination of both formats. The switch usually produces only one format. The user is not given a choice. The ASCII format offers the advantage of being directly printable for verification of the record contents produced. A binary format typically offers the advantage of encoding more information into fewer characters.

There are no official or de facto standards established for call detail record formats or the data contents. The current record is typically a common set of information elements with additional elements provided depending on the switch. The common set includes: the date and time the call was placed, the call duration, the called number, the extension placing the call, and the outbound circuit used. Additional elements include: an accounting or billing code, attendant or directly dialled indicator, identification of any shared switch hardware that was used, a reason why the call was attempted but not completed, and indicators that transmission of data occurred. If inbound calls are also recorded, then additional elements include: inbound circuit, caller identification by ANI and ISDN information.

The call detail interface described so far is a one-way flow from the switch to the call management system. The newer call management systems, which offer integrated switch management service, require a two-way interface to get information back into the switch. The information moves and changes the switch configuration database as made by the switch administrator, or as the result of an attendant making directory service changes. The information can also be status values and alarms input from sensors monitoring facility environmental and security conditions.

A very low protocol form may be used with one-direction interfaces. The call management system may connect as if it were a printer with only request to send (RTS), clear to send (CTS) or XON and XOFF flow control capability. With two-way interfaces, a more complex protocol, such as Bisync, SDLC or X.25 may be used.

Call management requirements for ACD systems are more complex than for PABXs, due to the real time critical operating conditions. Management information for control of the queuing functions within the ACD is typically provided by the ACD. Agent force scheduling and forecasting is typically handled by a separate force management system. The information required by the force management system is manually input or provided through a V.24 port from the ACD. Entering the information manually might be used when one force management system is shared by multiple ACD sites; however, this can be impractical as the size of the work force and number of ACD systems grow.

Depending on the ACD, multiple choices for interfaces to call detail information may be available. Examples are a V.24 port with one-way or two-way direction, with one or more protocol choices. Magnetic medium may also be available.

Call management information from an ACD can be extensive. Performance reports may be produced every 15, 30 or 60 minutes all day, every day. On very large ACD systems, the information can be for a thousand servers and proportionate number of trunk circuits, with call detail information for every inbound and outbound call produced at call rates up to 9 calls per second.

6.6 Call management trends

A clear trend is to merge call centre systems. Individual specialised systems which have addressed separate functions in the call centre are being merged into the same processing equipment providing call management reports. Modularity in the software, multi-tasking operating systems, and better adherence to OSI standards make this possible. Within the PABX environment, directory service and call accounting can be logically combined because they both need extension number and user name database information. The PABX equipment inventory and configuration information is another logical addition when the processing system is physically located near the PABX. Other additions can include property management systems, security monitoring, cable/wire management, reservations, message centres, hotel/motel features such as front-desk management, plus data network interfaces and reports.

Multi-site interface capabilities are becoming more commonplace as vendors attempt to provide more cost effective systems with new features. The call accounting systems will be able to join call detail records of calls passing through multiple switches for complete network path analysis and maintenance call tracing. With the evolution of more powerful processors, these capabilities will be extended to the PC-based systems.

On-line report modification and creation capabilities are being extended. Relational databases are being used with direct SOL query allowed. The user will be able to create special reports easily without returning to the vendor. User screens will be friendly and database information will be easy to load. Direct interfaces to spreadsheet programmes will be provided so special analysis can be performed and graphic display forms with colour selections can be used.

Availability and usage of ISDN will continue to expand, with a corresponding upgrade in systems to address the new requirements. Additional call information, hardware interfaces, circuit requirements forecasting, network control strategies, and maintenance specific information will be provided in reports and real time displays.

Better call accounting reports will be provided directly from the PABX and ACD systems. Shift scheduling and forecasting will be integrated into the ACD. Actual staff adherence to the shift schedules will be shown in real time by exception on supervisor screen displays and in system reports. These advances can be accomplished by improving the standard reports produced by the ACD and through marketing agreements which will allow integration of the add on system into the ACD. Such total integration provides a major benefit to the customer, since the add on call management system(s), procurement process, integration testing, and support issues can be eliminated as a result.

Strong competition among switch, software and service bureau vendors continues to result in new products and services. With the advent of the PC, the call management systems markets have been opened wide to many small entrepreneurial companies presenting new approaches. Requirements from call managers for more information presented more ways, even more quickly, and at lower cost are also contributing to the trends in call management. The continuing trend is to provide better and more cost effective call management systems and services to help people manage their call centres better.

6.7 References

Berman, C. (1990) Call Management: Moving Towards OSI Standards, *Communicate*, March.

Borthick, Sandra L. (1989) Old Habits Stifle Telemanagement Software Market, *Business Communications Review*, August.

Brown, Barbara, (1990) Paperless Call Accounting, *Teleconnect*, October.

Communicate (1993) No longer standing all alone, *Communicate*, May.

Communicate (1995) Logging cuts through the telecomms jungle, *Communicate*, April.

Costello, J. (1990) Counting call costs, *Communications*, October.

Costello, J. (1991) Applied call management, *Communications*, March.

Cummins, Michael and Stonebraker, Everett (1989) Total Quality Management of Telecommunications, *Business Communications Review*, December.

Farwell, C. (1991) ACDs in business, *Communications Networks*, November.

Fermazin, Tom (1991) Call Centre Managers: Rethinking Your Business Strategy to Take Advantage of Modern Tools, *Voice Processing Magazine*, January.

Fross, Alan (1988) The ACD Market Comes of Age, *Business Communications Review*, November-December.

Gordon, James. Klenke, Maggie and Camp, Kathey (1990) Centrex System Management Tools, *Business Communications Review*, April.

Gray, M. (1992) ACD is the ABC of service, *Communicate*, February.

Hayday, G. (1995) Beyond the call centre, *Business & Technology Magazine*, June.

Hayes, Robert and Harrison, Karen (1989) Chasing ACD Phantom Calls, *Business Communications Review*, September—October.

Horton, Venetia (1990) Call Management Stops Telephone Abusers, *Office Equipment News*, October.

Huffadine, Roger (1990) Trends in Call Management Technology, *Telecommunications*, December.

Johnstone, S. (1992) Users call the shots, *Communicate*, Februray.
MacPherson, Gordon (1989) How Staffing Affects ACD Trunking Requirements, *Business Communications Review,*, February.
Middleton, Peter (1989) Managing Your Calls, *Communications* , July.
Mikol, Thomas (1991) The IRS Collects With ACDs, *Inbound/Outbound Magazine*, April.
Sanchez, Robert (1988) Buffers and Borders: The Zenith Call Accounting Project, *Business Communications Review*, November—December.
Sinden, J. (1994) Network ACD enhanced information services, *Telecommunications*, May.
Stusser, Daniel (1989) Call Accounting: Not Just a Numbers Game, *Business Communications Review*, August.
Tucker, Tracey (1990) Call Accounting Update, *Teleconnect*, October.
Wilson, Kim (1991) Call Centre Automation: Enhancing TSRs with New Technologies, *Voice Processing Magazine* , February.

7. Voice processing

7.1 The voice processing market

7.1.1 Introduction

Voice processing is the term used for the technologies of storing and replaying speech, compressing and decompressing speech, the technologies of speech recognition and text to speech conversion (Gray, 1993; Scales, 1994; Communicate, 1995; Garlick, 1995).

While voice processing can be used in industrial process control and military applications to allow hands free operations of plant or equipment, the industry is increasingly focused on telephone callers accessing the voice processing equipment. What follows addresses these telephony oriented applications.

Voice processing products can be split into the following types:

1. Audiotex.
2. Voice response.
3. Voice mail.
4. Automated attendant.

Analysts value the European voice processing for 1994 at £300M. This clearly shows dramatic uptake in the European voice processing equipment over the next three years, and to a large extent mirrors the growth in the US that has occurred already.

7.1.2 Audiotex systems

Audiotex systems are used to publish spoken information to callers. Callers can interact with audiotex systems using DTMF tones, dial pulses, by speech power detection (grunt detection), or in more advanced systems by using Speech Recognition.

Many applications are currently available ranging from weather forecasts, games and contests, medical advice. These services can be accessed by the public using special telephone numbers. These numbers are tariffed at a premium over normal call rates; hence the term 'premium rate services'. The revenue from these calls is split between the PTT and the Audiotex service provider.

As an enhancement to basic Audiotex services, caller details can be recorded. This is used in games to identify the winner, or can be used for telemarketing purposes. A typical telemarketing use would be to automate the collection, collation and transcription of names and addresses of callers responding to a television advert.

Here the Audiotex system can be used to collect large numbers of caller details (e.g. names, address and telephone numbers). These details are then transcribed by data entry operators from the voice processing system into computer databases.

The benefit of collecting of names and addresses this way is that the peaks of traffic caused by media advertisements are accommodated 24 hours a day, 365 days a year and can be dealt with when convenient.

For example, a TV advertisement will generate a peak of traffic immediately it is shown. To reach the maximum audience, these adverts will be shown at peak viewing times in the evening, which is probably one of the most difficult times to employ staff to answer the calls.

7.1.3 Voice response

Voice response systems are used to allow callers access to databases using a telephone (Bailey, 1994; Moyes, 1995; Hooper, 1995). These systems require the callers to specify the data they are looking for, and so intrinsically are interactive services requiring the use of DTMF or voice recognition techniques. Historically these systems have used DTMF telephones, but increasingly in Europe are using speech recognition technology for caller interactions. The services offered by a voice response unit are usually limited to one or two and the voice storage capacity on disk as a consequence is very limited. The range of telephone line sizes of voice response units commonly found is between 4 and 30 lines.

Traditionally the voice response system interfaces to the computer containing the database by emulating a terminal. A common form of terminal emulation used is IBM 3274 / 3278 SNA. The voice service 'reads' data from the terminal emulation 'screen' and speaks this information to the caller. The caller can select the information to be spoken and the screen being used. The DTMF commands sent by the caller are translated into 'key pushes' on the emulated keyboard.

Other systems use directly interpreted data, especially formatted for the application. This data is sent by any of the traditional data communications protocols. However X.25 is a particularly popular technique due to the flexible nature of the data communications network and the software support in mainframe computers. Typically used in home banking applications, this technique allows simpler systems to be realised.

More advanced systems in the IBM system environment use LU 6.2 and distributed database enquiries based on Standard Query Language (SQL) techniques.

Voice response systems have to readily speak numbers to callers. In order to facilitate this, and because of the limited vocabulary needed, the speech may be stored in the voice processing systems semiconductor memory (RAM). The words are brought out of memory by using a microprocessor and presented to digital to analogue converters. This reduces the load on the system compared with taking such speech from disk and hence makes the concatenation of numbers easier to achieve without recourse to specialist disk controller hardware and software.

As an adjunct to this feature, some voice response systems have text-to-speech capability. For example this is used to speak item descriptions to callers from a stock list. It is considered impractical to individually record each part description when there may be many thousands, or even millions of distinct descriptions.

Whilst not directly being part of a voice processing system there is a growing use of fax response systems. These allow the caller to select information using voice processing techniques and then send out a fax of the required information to the caller.

In many cases it would be difficult to speak the information to callers, and there is considerable benefit in a fax message being sent.

7.1.4 Fax responses

Fax response systems, when incorporated within a voice processing system, have two basic modes of operation; store and forward on demand and the on line transmission of information constructed for the particular caller from a database. The first mode is used to make a standard document, such as a map, available to a large number of callers. The second is used to allow caller relevant details to be put in to a fax and be sent. Typical applications include faxing of bank statements, inventory status etc.

The fax information sent is chosen during a normal telephone call to the voice processing system. During this call their fax machine number is entered by the caller. When the desired information has been selected, the system automatically dials up the identified fax machine and sends the information.

The problem with this approach is that the owner of the system pays for the call. It is not surprising therefore that many of the applications of this technology involve closed user groups (such as bank account holders), which allows a charge to be readily levied for this service.

However, according to industry sources approximately 80% of the installed base of fax machines support the reverse polling mode of operation. Here the fax response system would simulate a fax machine already loaded with a document. Callers could then make their fax dial up the system and it would send them the documents. This feature is not widely used at present.

7.1.5 Voice Mail

Voice mail was one of the first voice processing applications developed during the late 1960s. Voice mail, as the name suggests, allows callers to leave voice messages for other users of the system

(Fennel, 1993; VI, 1995). Typically connected to a PABX via a few analogue telephony interfaces, these systems are widely used in North America where the large penetration of DTMF phones facilitates its use.

The voice messages are usually short reminders or enquiries. The person intended to receive the message is notified by a message waiting lamp on his/her telephone, or by a different dial tone when the recipient uses their phone. Sometimes a radio pager is used to alert the recipient of new voice mail.

Inexpensive telephones with message waiting lamps are not as readily available in Europe as they are in the USA. The other alerting systems are generally thought of as unsatisfactory or too expensive for widespread use.

One popular use of voice mail is as an adjunct cellular telephones. Since cellular phones may be switched off for substantial periods the use of a voice mail box offers many advantages in business situations.

Within the UK voice mail competes directly with telephone answering machines and has not been as successful as in North America.

With the advent of GSN cellular telephones and the concept of a personal number for life associated with GSN, the need for large call delivery oriented voice mail systems will increase and probably become an accepted facet of business life within the European business community. The Pan-European GSN standard will considerably facilitate the take up of these services.

Voice mail systems can be stand alone or units can be networked to provide a seamless but geographically separated system. The interface protocols between voice mail systems have been proprietary until recently. Now there are the AMIS standards for voice mail systems. One is a digital standard based on the ITU-T recommendation X.400, which is a wide ranging messaging standard based on OSI principles. The other is an analogue standard using DTMF control tones associated with the analogue transmission of the voice files. The widespread adoption of these standards is not happening as rapidly as was expected.

The X.400 system is split into two main parts; the User Agent (UA) and the Message Transfer Agent (MTA). On top of these two standard parts is the particular manufacturer's user interface. The user interface interacts with the UA to set parameters such as destination address. Once the necessary parts are complete, the message and UA details are passed to the MTA for delivery to the destination. This can be by a number of data communications protocols such as X.25, but this detail is hidden and made irrelevant to the user. X.400 is currently used by electronic mail systems. A number of voice mail suppliers have it as an available option.

The destination address can be selected using a ITU-T recommendation X.500 directory service. This can allow users to select destinations using database search techniques. These standards will potentially allow different manufacturers' systems to inter-work and users to be able to use just one set of command structures.

However, the transmission of messages between voice mail systems by these techniques is at best only moderately faster than real time. For this reason geographically distributed voice mail systems are generally used only where necessary, such as in the transatlantic situation. The great majority of voice mail systems at present are not distributed, but work by having users connect to them using private or public telephone facilities. In the US the majority of sales are for systems of 4 lines upwards.

There are many features in the currently available voice mail systems. They use DTMF tones to control the addressing and selection of menu options. DTMF tones are also used to identify the caller. This means that usually only business users in the UK can take advantage of these systems.

However a limited number of systems support speaker dependent recognition technology to receive commands from callers. This is very suited to voice mail systems since inherently the users have to identify themselves to the system initially by using a password. The speech recognition templates for that particular caller can then be recalled and used. Command words can therefore be chosen to suit customers on an individual basis.

One important aspect of a voice mail system is the management of the users, mail and system. When selecting a voice mail system these features should play an important part of the decision.

Guidelines for the size of a voice mail system are difficult to define, since the number of lines and disk size depends on the usage of the system by users. However, for example, in a medium use environment one voice mail system port for 30 users and a 3 minute voice storage capacity per user could be a reasonable starting point when sizing the system.

7.1.6 Automated attendant

This voice processing application is usually used as an overflow mechanism in conjunction with switchboard operators. In North America, automated attendants are used with the callers DTMF phone to reach extensions directly (so called simulated DDI). The caller is prompted for the required extension or department by the system. If the person being called is busy or does not answer, a voice mail application is usually made available to the caller so that an appropriate message can be left.

The use of automated call handling is becoming important for large corporations striving to reduce costs, make their services available 24 hours a day and improve their image. Here the concept is to handle caller queries as comprehensively as possible and not involve an operator if at all possible. Customer care is an important public relations issue. There is a reluctance to accept this technology in Europe because of the perception of talking to a machine as being undesirable. Since caller sessions can be more interactive than in the case of talking to an answering machine, this barrier is being overcome.

Links to databases and diversion to live operators can be used to enhance the capabilities and usefulness of automated attendant.

7.1.6.1 *Links to PABX/ADC*

In many cases the voice processing system is installed behind a PABX or ACD. This allows calls to be handled automatically or by live operators. Callers can be readily transferred from voice processing system to live operator or vice versa. However, in many situations the maximum benefit can be gained by having the caller screened by the voice processing system first; details obtained and then the caller being transferred to a live operator only if necessary. The details obtained by the voice processing system are shown on the operators VDU prior to the call being transferred.

There are many problems associated with achieving this functionality, centred around the lack of integration between PABX/ACD, host database and voice processing system. In an ACD for example the call is presented to a number of operators. The first operator to accept it gets the call. The information about which operator took the call may not be available to the host computer to present the relevant details on the correct VDU. Also the call may be put through before the VDU screen is presented. There is a wide variation in ACD/PABX functionality between manufacturers.

While there are many proprietary standards available from individual manufacturers there is no clear de facto standard emerging in this area of Computer Supported Telephony Applications (CSTA). ECMA and ANSI standards are being formulated but are not available yet.

7.2 A voice processing system

In order to understand some of the issues involved in the many applications of voice processing let us examine a generic voice processing system (Farren, 1994; Moscom, 1995). This description is not intended to relate to any particular manufacturer or model.

Figure 7.1 is intended to depict a general voice processing system. In the real world keen attention is payed to the number of lines supported, the amount of disk storage available and the speech recognition capabilities of the system. Architecture, software and hardware, and modularity are also key issues.

The basic subsystems of the voice processing system are linked by two buses. The speech bus conveys caller speech to the various speech processing elements and voice from the system to the caller. In advanced systems a degree of configuration of these speech paths is possible using digital cross connection switches.

The computer bus links the subsystems to the voice service applications processor for monitoring and control purposes. This bus is also used to deliver speech information to and from the voice com-

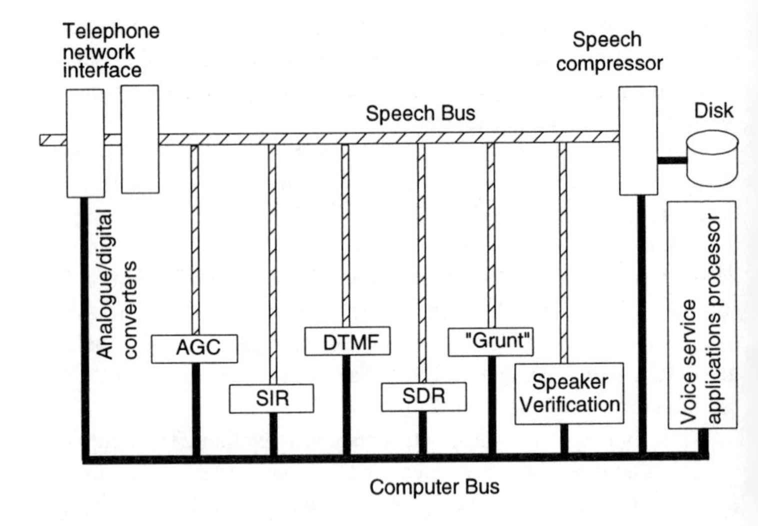

Figure 7.1 Generic voice processing system architecture

pression/ decompression module. If the subsystems are intelligent, then they are down loaded with software using this bus.

The system functions by recording and replaying compressed speech from disk under the control of the voice service applications processor. There are two basic architectures; those based on PCs and those that are proprietary. PC based products in general address the low end of the market in terms of line size and usually focus on just one application area; proprietary architectured products on the other hardware suited to more general purpose use with higher performance and more line ports and disk storage. There are many PC based products in the market place. One of the benefits claimed for the PC products is the 'openness' of the platform. This openness is usually only effectively available to companies with experienced programmers. Also, because of the limitations of the platform for voice processing applications, it may be difficult to incorporate a wide range of features and applications in the same PC.

The callers access the system via the telephone network interfaces. There are two main types of telephone network connection: analogue and digital. The trend is increasingly towards digital network connection. This is because of the improved quality, the convenience of the physical network connection and the increased functionality and reliability possible.

7.2.1 Analogue network connections

There are four types of analogue network connections: direct dial inwards, normal telephone, Centrex and E&M.

7.2.1.1 *Direct Dial Inwards (DDI)*

This type is widely used by the Audiotex service providers and is currently offered in the UK by BT using the Derived Services Network (DSN) and Mercury using their normal network. The BT DSN has been dimensioned to cope with the large volumes of traffic that can be associated with Audiotex applications. The telephone network forwards the last 2 or 3 digits dialled by the caller to the voice processing system. These digits select the opening service given to the caller. This circuit type only supports incoming calls. It has been used with PABX systems in the past to allow callers to dial straight into a particular extension telephone on the PABX from the public telephone network.

The service access digits are sent by decadic dial pulsing from the telephone exchange to the voice processing system. When the digits have been completed (detected by a quiet interval significantly longer than the inter-digit pause) the voice processing system validates the access code and returns number unobtainable, engaged or ringing tone to the caller, depending on the availability of the service.

Two or three bursts of ringing tone are given before the caller is connected to the selected voice service. (In a PABX environment the ringing would continue until the extension was answered.)

At this point call charging is initiated by the voice processing system reversing the polarity of the battery voltage it supplies to the telephone exchange. This means that call charging is initiated by the equipment, not the telephone network. As can be imagined this results in the rigorous testing of the equipment before it is approved for connection to the public network.

7.2.1.2 *Normal telephone type*

The voice processing system mimics a normal telephone. This kind of connection is used for information update lines or when the voice processing system is placed behind a PABX. The telephone exchange sends ringing current cadences to the voice processing system. After two or three bursts the voice processing system answers the call and thus stops the ringing current from the telephone exchange by looping the circuit. The caller is also then connected through to the voice service. Since there is no pre-selection of the service (cf. DDI), the caller may have to interact with the voice processing system to then select the service required. Caller charging commences as soon as the voice processing system answers the call.

The call may be terminated by the exchange temporarily removing the battery from the circuit, or in some exchanges by the voice processing system un-looping the line. This type of connection can support both incoming and outgoing calls.

There is a problem for this kind of connection associated with detecting the release of the call by the calling party. There may be no firm indication to the equipment that the caller has released from the telephone network. For example, some PABX systems may give dial tone when the caller releases, some may give silence.

Whereas normal human-to-human conversations naturally terminate in the face of silence, human machine conversations do not. Unless the machine has a clear indication from the telephone network that the caller has gone, it will continue trying to interact with the caller. The procedure usually adopted to overcome this is for the voice processing system to give short fixed length messages, or for it to prompt the caller for an interaction (DTMF, Grunt etc). If the caller does not reply the system drops the call so as to be ready for new callers. Alternatively, if the network / PABX returns dial tone to the voice processing system when the caller hangs up, a dial tone detector can be used. However to be effective the dial tone detector must reliably detect dial tone in the face of the voice processing system replaying speech. Digital signal processors are suitable for the filtering and detection function in this case.

7.2.1.3 *Centrex type*

This is a variant of the normal telephone type. The more popular large capacity PABXs support the connection of voice mail systems through interfaces of this kind. Whilst not being described as Centrex, the interface is very similar. These interfaces differ from normal extension ports in that DTMF tones identifying the calling party (if it is an internal call) are sent by the PABX as soon as the call is answered by the voice processing system. Centrex features offered by the telephone exchange are intended to replace those offered by on-site PABXs.

Service selection digits (dialled by the caller) are sent immediately after the call is answered by the voice processing system from the telephone exchange. As in the case of DDI, these digits correspond to the last 2 or 3 digits dialled by the caller. However, these are sent by DTMF tones to the voice processing system.

Once the call is established a recall indication can be sent from the voice processing system to the exchange. This takes the form of a timed loop break (100ms) or an earth applied to one leg of the telephone cable pair.

The telephone exchange then sends a dial tone to the voice processing system as an indication that it can dial out the number using DTMF tones and invoke features such as three party conference calls.

When the original calling party or the voice processing system terminates the call all parties are usually released from the connection.

7.2.1.4 *E&M type*

Sometimes referred to as DC5 or DC10 signalling systems the Ear and Mouth telephone interfaces are used by VPSs when connecting to ACDs or PABXs.

There are many types of E&M interfaces, depending on region and PABX. E&M can be configured to support both incoming and/or outgoing calls. The E&M interface is designed to avoid incoming and outgoing call collisions by operating a simple signalling protocol. These interfaces are falling out of use in voice processing applications.

This is because in many applications it is desirable to be able to connect the caller to a live operator. To do this using E&M interfaces would require the voice processing system to route the caller out on another circuit back to the PABX (so-called tromboning).

The voice processing systems call handling capacity is reduced because it is associated with the call for its complete duration. Also, twice the numbers of ports are required both on the PABX and the voice processing system compared with using normal or Centrex type PABX ports equipped with the recall and dial out feature.

Using the voice processing system in this way may make it "Call Routing Apparatus" in regulatory terms, adding significantly to the regulatory approval issues and to installation and on going support standards. Most voice processing equipment manufacturers have therefore not taken the step of becoming call routing apparatus.

7.2.2 Digital network connection

There are two main standards for digital connections: T1 (24 channels) in the US and Japan and E1 (30 channels) in the rest of the world. T1 standards are set by ANSI; E1 standards are set by CEPT. T1 has a clock rate of 1.544Mbit/s and E1 2.048Mbit/s. T1 signalling is carried within each telephone channel; E1 signalling is carried in time slot 16.

7.2.2.1 *Channel Associated Signalling (CAS)*

In the E1 version of CAS the E1 framing structure is used to transmit 4 bits of information per channel during 16 E1 frames. For historic reasons the 4 bits are termed the A,B,C and D bits. Each E1 frame consists of 32 timeslots of which 30 can be used to carry one voice conversation each. Timeslot 0 and 16 in the E1 link are reserved for signalling use.

Two channels' information is sent in each E1 frame in timeslot 16. Timeslot 0 is used for other purposes such as frame synchronisation signalling. The sixteen E1 frames are termed a superframe. Each channel's signalling bits are pre-allocated into a particular frame in this scheme. Whether it is necessary or not, new signalling bits are sent in each superframe for each channel.

The signalling bit rate for each timeslot is 4kbit/s, which is adequate for most telephony applications. The bits are used to mimic line states, such as battery presented etc. and dialled digits. The dialled digits are represented by simply sampling the dialling waveform and transmitting its state in the relevant bit and frame within the superframe.

In the T1 case a similar system is used, but the bits are hidden in the caller timeslot bits. Every 6th frame, bit 8 of the caller timeslot is replaced by a signalling bit. This is termed robbed bit signalling. A multi-frame structure is identified by the use of a spare bit at the end of each T1 frame (bit 193). By this means up to 4 (but usually 2, the A and B bits) bits per channel are transmitted. The bits are used in a similar way to the E1 case. There are many variants of the use of the CAS bits.

7.2.2.2 *Common Channel Signalling (CCS)*

In common channel signalling systems a more complex and flexible protocol is used. These systems are not at all common in the T1 system because robbed bit signalling is unsuitable for this application. In E1 systems timeslot 16 in the E1 frame is used to carry HDLC formatted data. Any caller timeslot of the frame can have signalling data inserted in timeslot 16 at any time. There is no pre-allocation of bits within the superframe to timeslots. Signalling information is only sent when necessary.

When primary rate ISDN protocols become available they will use a CCS system as described above. Currently DPNSS and DASS2 CCS schemes are widely used in the UK. DASS2 is currently the only way to interface at E1 rates to the BT public Network. Mercury supports DASS2 and a CAS scheme at the moment for public network access at E1 rates.

7.2.3 Signalling information extraction/insertion

Usually in voice processing systems signalling information sent from the telephone exchange is extracted to select the opening service. Information can be sent by decadic dial pulsing loop disconnect signalling or DTMF tone and/or digital signalling in E1 or T1 connections.

In more advanced E1 and T1 systems, signalling information can be sent as HDLC framed data such as in ISDN (ITU-T), DMI(AT&T), DASS2 and DPNSS signalling schemes. These allow the calling party to be identified if this information is provided by the PTT. In the USA this feature, known as Automatic Number Identification (ANI) is widely supported.

Most current US systems currently use DTMF techniques to receive this data from the network. In the UK regulatory and legislative issues may prevent its widespread availability. The feature of providing this information within the UK is known as Calling Line Identity (CLI).

Outbound signalling can be performed using a new circuit or the same circuit used for the incoming call once the exchange has been notified by a recall signal, by any of the above methods.

7.2.4 Analogue to digital conversion

In the case of analogue network connections the caller's speech needs to be digitised before it can be used. In the case of digital connections the digitalisation process is usually carried out in the callers local telephone exchange. A-Law (European) and μ-Law (USA and Japan) are companding rules used to enhance signal quality.

Companding is the term used to describe the non-linear allocation of the 256 levels in the 8 bit timeslot sample to analogue signal levels. Usually the digitalisation process creates 8 bit quantised values. This gives adequate audio quality for telephone network use.

Either before or after the digitisation process the caller's speech is processed to detect tones, to amplify it to a reference level etc. (e.g. DTMF tone detection, automatic gain control, etc.).

In advanced systems these processes are carried out by Digital Signal Processes (DSPs). These are special purpose microprocessors with an instruction set specially for these purposes. In early voice processing systems these functions were carried out in the analogue domain using op-amps etc. These, by comparison, are difficult to manufacture and modify in service.

Analogue systems using tightly toleranced components also have a tendency to degrade with time due to component ageing and are inflexible, difficult to modify and perform field upgrades. On the other hand DSP software can be altered at will to modify and enhance functionality and no degradation over time is usually experienced.

DSPs use the digital representations of analogue waveforms obtained from analogue to digital conversion and manipulate them using software. All the usual analogue filtering techniques can be realised and many more, such as adaptive filtering, correlation and convolution, delay lines, matched filtering, Fourier analysis and echo cancellation.

7.2.5 Speech compression

In order to maximise the effective size of the speech storage in the voice processing system caller prompts and inputs are usually compressed before storage and decompressed during replay.

Traditionally achieved by using many simple special purpose chips, the speech compression technology has gone full circle. DSPs can be and are used for compression algorithms. However, the expense and bulk of this approach for voice processing systems with many lines has caused the introduction of special ADPCM chipsets which can compress/decompress 60 channels of speech into 30 channels space.

These chipsets are widely used in telecommunications transmission equipment for compression purposes.

Currently Continually Variable Slope Digitalisation (CVSD) and Adaptive Differential Pulse Code Modulation are the preferred methods. CVSD is used at a rate between 20kbit/s and 32kbit/s, ADPCM at 24kbit/s or 32kbit/s (i.e. 3:8 or 2:1 compression).

ADPCM at 32kbit/s introduces negligible distortions and is generally considered superior to CVSD, although it is more complex and expensive to realise. ADPCM has become the preferred compression technique.

More advanced compression algorithms yielding 8:1 compression ratios are available, but either the quality is poor or the hardware cost and complexity is too great for application in multi-channel voice processing system.

7.2.6 Text to speech conversion

This capability is used to convert computer based text (held in a database or disk files) into speech for transmission to the caller. Normally the caller hears pre-recorded and replayed speech. However, using this technique to speak the contents of a database may be impractical because of the huge repertoire involved.

There are two main parts of the text to speech conversion process. The text is analysed and converted to phonetic equivalents. The phonetic equivalents are then spoken to the caller by a formant voice synthesiser. This device mimics the resonant cavities of the human vocal tract by using variable filters (usually realised in DSPs) and simple impulse excitation of these filters.

To complete this filter arrangement the fricative (hissing) part of the speech output is formed from a filtered and amplitude controlled noise generator.

There are usually five formant filters which model the human voice tract and as few as three can be varied and still achieve reasonable speech quality.

The process of translating from text into a phonetic representation is well developed. Some text-to-speech converters are capable of pronouncing over 100,000 words correctly, and can use the context of the words to derive the correct pronunciation.

Subsequent prosodic processing of these phonetic representations, introduces the intonations, inflexions and the rise and fall of the volume of the speech to make the output more acceptable.

However, to date the output quality achieved by this process still has a pronounced robot like quality, and as a result this capability is not widely substituted for replaying pre-recorded speech.

7.2.7 Speech recognition

It is this capability that will alleviate the current lack of DTMF telephones in Europe and allow voice processing technology to reach its full potential. There are two main variants; Speaker Dependent Recognition (SDR) and Speaker Independent Recognition (SIR).

Speaker dependent recognition systems need training to individual callers' voices, whereas speaker independent systems do not. Because of this SDR systems are falling out of favour due to the training chore, even taking into account the considerably increased complexity and cost of SIR systems.

Speaker independent recognition systems analyse the caller's speech and, in parallel for all the available words, match certain qualities of the speech against templates corresponding to the words to be recognised.

These templates are generated by taking many samples of the words desired from the range of accents likely to be encountered. These samples are usually processed off-line to create the templates. Typically between 100 and 200 different speakers are required to construct the templates.

The quality of speech examined is usually the frequency content of the spoken word. This extracted information typically corresponds to the syllables of the spoken words.

Commonly only isolated words can be recognised. However, there are other speech recognition products being introduced capable of connected word recognition. This can separate and recognise a series of words spoken without significant pauses. For example, connected word recognition systems allow the caller to speak a string of digits to the system. The speech recognition system then has to separate the individual digits in order to recognise them.

Even if the SIR system has a recognition performance in the high 90% region individual digits, the length of string that should be attempted is practically limited by the overall recognition accuracy that may be achieved. For example, if a recognition accuracy greater than 80% is desired and the single word performance is 95%, then a 4 digit string would have a 81% probability of being accurate.

It is also quite difficult to identify which spoken digit is incorrect and then prompt for that digit's re-utterance by the caller. For these reasons connected digit recognition systems are usually programmed to prompt the caller to speak a large number a few (3 or 4) digits at a time.

7.2.8 Word spotting

Word spotting systems continually examine callers' speech in an attempt to recognise an embedded word. This approach has the advantage that it may be able to find one of the words in an unstructured reply by the caller to a prompt. For example, the system might prompt the caller to say 'yes' or 'no'. If the caller replies 'Well, yes, then'. a normal speech recogniser would not find the embedded 'yes', whereas a word spotting system would.

7.2.9 Speaker verification (voice printing)

This is a system where the voice print of a caller is taken in an enrolment session and subsequently used to authenticate the caller's identity. The caller is prompted to speak a word or phrase which is then compared with the samples taken during the enrolment session.

Speaker verification system performance is classified in two ways. The percentage of false acceptances and the percentage of false rejections. The system is usually biased to reject rather than accept people. Systems with a performance of a few per cent false rejections, and a fraction of a per cent false acceptances are available. Obviously any change in the caller's voice due to a cold or stress can increase the false rejections made.

When perfected this feature will enable many over the telephone financial transactions to be completed securely. It is envisaged that these systems will be used in conjunction with a limited number of live operators who will help people who have failed the security check.

7.3 Future directions

There is an continuing trend towards the miniaturisation of equipment; increasing amounts of voice storage and easy to use voice service programming tools. To fully reach the potential that this technology has to reduce costs, it will need to be more closely integrated with the corporate I.T. infrastructure. The technology of voice processing, through advanced system software, needs to be made more available to end users.

One of the key technology breakthroughs will be comprehensive speech recognition technology capable of recognising many thousands of words and also capable of understanding the contextual issues within speech. The ability to recognise a few words is with us today. The incremental improvement to recognise a few tens of words is in sight. However, the step of being able to interact with the machine as if with a human is not clearly visible (except on Star Trek!).

In order to enable voice processing to leave behind the stigma of the answering machine and reach its full potential, the ability to hold natural and non-deterministic dialogues is key. Simply programming the machine with a tree of responses and menus is not good enough. The human-machine dialogue should be natural, allow interruptions and leave people with some doubt as to whether or not it is a machine they are talking to. This may well need advances in artificial intelligence beyond current technology. Some how the machine must be programmed with an 'opinion' so that the thread of the conversation with the caller can be maintained. The future for voice processing technology is bright. The ability to use the telephone as a cheap terminal device, able to access data and allowing callers to command and control systems has not yet been fully explored. The challenge to go beyond the current talking computers can and will be met.

7.4 References

Bailey, C. (1994) Computing over the telephone, *Voice International*, October/November.

Communicate (1995) A final solution to agonising hold music, *Communicate*, February.

Farren, P. (1994) Voice processing: the PC takes over? *Telecommunications*, May.

Fennel, J.K. (1993) Voice processing on the mobile network, *Telecommunications*, February.

Frost & Sullivan (1990) *The European Market for Voice Processing Systems*, Frost & Sullivan Ltd., London.

Garlick, S. (1995) Speech technologies meet the needs, *Voice International*, April/May.

Gray, M. (1993) Voice processing gets its open hearing, *Communications International*, November.

Hooper, N. (1995) Call management, messaging and IVR equipment, *Voice International*, February/March 1995.

Moscom (1995) Speech recognition technology in telecommunications, *Voice International*, December/January 1995.

Moyes, I. (1995) Spreading the word, *Voice International*, February/March 1995.

Scales, I. (1994) Opportunity knocks, *Communications International*, May.

Tetschner, W. (1991) *Voice Processing*, Artec House, Norwood, ISBN 089006 468 7.

VI (1995) From hotdesking to helping the homeless, voicemail evolves, *Voice International*, June/July.

Walters, R.E. (1991) *Voice Information Systems*, NCC, Oxford, ISBN 1 85554 075 4.

8. Electronic data interchange

8.1 Fundamentals of EDI

Electronic Data Interchange (EDI) is a subset of the messaging marketplace. A working definition of EDI is the computer to computer transmission of documents using agreed formats and protocols.

It is rarely mentioned but often assumed that EDI refers to interchanges between businesses (Houldsworth, 1990; Young, 1993). Whilst this may be largely true, in practice intra company exchanges are not precluded and large companies may see many benefits in adopting EDI standards for internal use.

There are two good reasons to introduce EDI into a business: firstly to reduce the overheads of running a business and secondly to enable the provision of improved service to customers.

The tangible benefits of EDI include:

1. Less manual data entry leading to an increased accuracy.
2. Reduced clerical overheads resulting in a lower cost base.
3. Faster delivery of data allowing shorter lead times and even enabling lower stock levels with consequent cost savings.

8.2 EDI in the messaging marketplace

The marketplace for EDI goods and services has been operating for some time but it is only since November 1990 that a complete OSI standard for moving EDI messages over the OSI stack has been fully ratified. To understand how this standard, X.435, was arrived at a brief history of the messaging marketplace is included with an outline of the X.400 standard used for inter personal messaging (IPM).

The area of messaging has been the most active area for standards work over the past couple of years. This may well be because it takes about six years from the recognition of a business need to the provision of standards based products to fit the requirement. The time lag is due to the four year cycle of the international standards body, the ITU-T, and then the two years it may take to develop and bring a standards conformant product to market.

The messaging marketplace used to have a black and white split into EDI and Electronic Mail, however recent trends are creating many shades of grey and it is the standards process which is allowing a much richer range of integrated services to become available from the major service providers (Scales, 1993; O'Brien, 1994; PM, 1995; Hunt, 1995). As the marketplace becomes more sophisticated the ability to deliver a range of messaging products down a single pipeline, thus simplifying the underlying network, is becoming a key differentiator in the messaging marketplace.

8.3 Messaging standards

Much standards work has previously been done in the messaging arena, however the standards fell into two broad categories which may be summarised as follows:

1. The 'movement' standards, building on the previous work which established the OSI seven layer model and defined the lower levels up to transport (Level 4), these now take the model to the application layer, Level 7.
2. The 'message' standards, work done to define the content, layout and data sets which are then moved over the above. This is the traditional realm of EDI and leads to definitions of EDI such as 'The use of computers to exchange documents using agreed formats and protocols'. Examples of the 'movement' standards relating to messaging are X.400, FTAM and OFTP (Communicate, 1994a). These standards place the minimum of constraints on the content of the message but concentrate on the delivery mechanisms.

Whilst X.400 and FTAM have been driven centrally by the ITU-T, message standards have tended to be community specific. The message standards, such as the American ANSI X12 and EDIFACT consist of two parts (Dunlap, 1993; Communicate, 1994b):

1. A language or rule set for constructing an interchange,
2. Building blocks; a hierarchy of parts to which the rules are applied to build up an interchange.

The building blocks start with data elements, which map onto familiar data items such as part number.

These are combined to form segments which are groups of related elements e.g. name and address. Some syntaxes allow compound elements and compound segments.

Segments are used to build EDI messages (also known as transaction sets in ANSI X12). A message is analogous to a document such as a purchase order.

Finally messages are grouped together in batches for transmission and this is known as an interchange.

Interchanges for all the main EDI standards include the concept of a header containing special segments to pass control information about the sender, recipient, date/time and a reference number.

Whilst many data elements are alike in the three main EDI standards (EDIFACT, ANSI X12, TRADACOMS), the main differences lie in the characters used to separate fields, the definitions of segments and messages.

As mentioned above the process of defining specific messages has been community, even industry, driven. So whilst EDI messages may conform to an international standard such as Edifact for the syntax of the message the actual data set used is often industry specific.

The situation is further confused in as much as there are several national and industry standards bodies, e.g. ANSI, SMMT, who have defined widely implemented standards ahead of the ISO/UNEDIFACT work becoming internationally adopted. Worse yet much of the EDI marketplace relies on proprietary communications rather than OSI.

Recently however the two strands have begun to converge with the emergence of standards which relate to both, messages and movement and also addressing i.e. X.435 and X.500. A major current activity of the service providers, such as AT&T EasyLink, is offering a bridge between the older and the newer systems so that previous investment can be protected but access and connectivity can be provided to those new entrants to the community wishing to conform to the latest standards.

This explains the apparent paradox between a six year lag after standards are formed and there being an established EDI community. A six year wait is clearly unacceptable if a business edge is sought so a means of making new technology more immediately available is clearly required.

To achieve such a target it is necessary to distinguish between the principles involved and the actual protocols.

Historically the U.K. EDI marketplace has been dominated by the use of VANS with a high degree of added value on the network. The added value has taken the form of verification of the data set, security and trading relationship checking, translation services between data sets and to some extent communications protocol conversion.

The communications prevalent have been proprietary or at best de facto, either IBM protocols or OFTP (albeit over X.25). Whilst the Edifact syntax is standardised the number of registered international standard messages is relatively small. Most recognised messages relate to smaller industry groups and are awaiting international ratification.

To fully understand the emergence of OSI standards for EDI it is necessary to take a brief look at the electronic mail marketplace.

E-mail has had one characteristic which has made a significant difference to its development. Whilst EDI is primarily used for inter company communication E-mail has tended to be intra company. Therefore the early E-mail marketplace showed a host of proprietary systems, often incompatible. As companies sought to combine internal systems a standard was required and X.400 was born.

The U.S. EDI marketplace initially followed a different standard to Europe, ANSI X12 instead of EDIFACT. However a larger number of messages were defined in ANSI X12 and the existence of these messages enabled a different solution. More generalised premises based software was possible and the marketplace which evolved reflected this. The value added by the network was less, the off net processing correspondingly higher.

However for E-mail there was little difference in the standards U.K. to U.S. so the same profile evolved, i.e. a large number of proprietary systems. The market for E-mail on both continents is however a little further advanced than EDI in as much as E-mail was seen as the ideal vehicle to further the completion of the OSI model, perhaps as it was a fairly easy concept to grasp.

So the second wave of E-mail products were able to convert the principles of the proprietary systems into the 1984 X.400 protocols to give an OSI solution to the E-mail requirement. This is now being refined to the 1988 X.400 model which gives a much closer match to the 7 layer model. This is because it splits the application down and actually uses layers 5 and 6 properly instead of being a complete application (layers 5, 6 and 7 in one) sitting on layer 4 as the 1984 implementation did.

The U.K. EDI marketplace, as the E-mail before it, is now taking advantage of the second generation of thinking. Premises based software is now being produced to move the processing off net, with the resulting requirement for a simpler value added service and a corresponding cheaper and simpler pricing structure from the service providers (Figures 8.1 and 8.2)

The network also provides extra delivery mechanisms, both in terms of the devices which may be used and the protocols used to reach them.

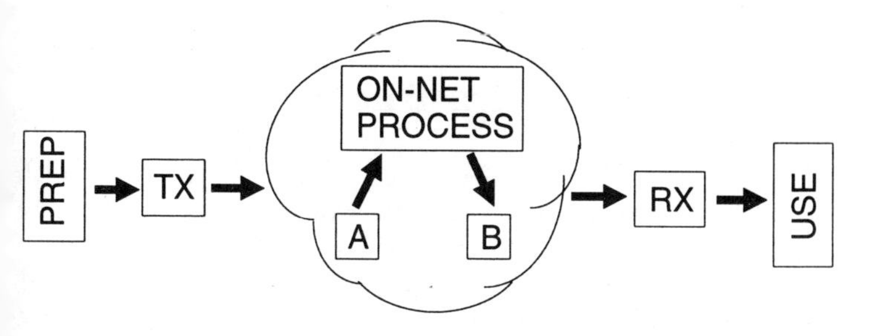

Figure 8.1 Network value added

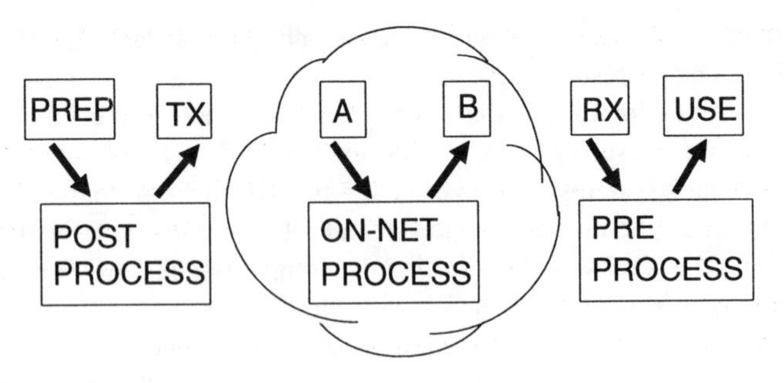

Figure 8.2 Premises based processing

These changes enable smaller players to enter the market, extending the accessibility of EDI to the next tranche of customers who were previously priced out.

The connectivity, translation and other services are achieved by means of gateways on the service. Each gateway provides a specific service. This means that a new service can be easily added or revised so that as a standard emerges in a particular area the gateway can be enhanced to meet the protocol requirements without changing the principle behind the service offering.

VANS will continue to play a key role in the transition from proprietary network protocols to OSI transport through their translation facilities. By offering translation it releases the individual company from the need to harmonise its network offerings and change procedures to those of all its trading partners.

8.4 Technical infrastructure

X.400 is generally used to refer to a series of recommendations which provide an international standard for structuring and transmitting electronic mail messages. So far there have been two iterations of the four year cycle, yielding the 1984 and 1988 version of the standards.

Since the early days it has been recognised that the base standards were suitable for transmitting other types of messages other than IPM so the 1990 standard for P_{edi} built on the earlier work to define a complete ISO stack for EDI. However, industry could not wait for the

1990 standard so work was done in America and Europe to adapt the 1984 X.400 standard for EDI.

The key elements of the 1984 standard are the message transfer agent (MTA), the user agent (UA). These are processes (composed of software and hardware) which provide services for moving messages from site to site and enabling users to send and receive messages. Figure 8.3 illustrates these concepts. Note that human users are explicitly included in the model!

There are other key concepts in X.400 as follows:

1. MHS — message handling system
2. ADMD — administrative domain, a public service provider

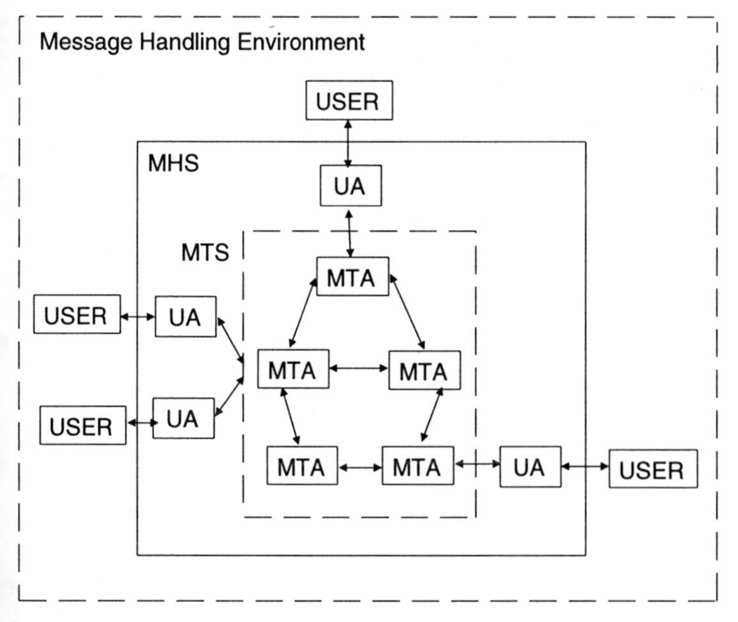

Figure 8.3 Functional view of MHS model

3. PRMD — private domain, a company's internal service
4. ORName — the originator/recipient name, a specific form of address required by X.400
5. P1 — the protocol between MTAs, equivalent to the envelope
6. P2 — the protocol between UAs for IPM, equivalent to the structure of a memo with header and body
7. DNs — delivery notifications, confirmation of receipt

As well as some refinements and extensions to the 1984 standard (e.g. P2 was enhanced and became P22), another key concept was introduced in 1988, that of the message store (MS). A MS allows a MTA to store a message it cannot immediately deliver to a UA e.g. when the UA is implemented on a PC which is periodically turned off.

Whilst in practice the MS is extremely useful its use is optional. For the sake of clarity in the following discussion the MS is conveniently omitted.

An X.400 message consists of an envelope (P1) and content (P2 for IPM). A P2 content has a header and a body, as in Figure 8.4. In adapting X.400 1984 for EDI two approaches were possible. The first, adopted in Europe on the recommendation of the CEC, is to put the EDI message into the P2 body part. A copy of the originator and recipient from the EDI message header are put into the P2 heading.

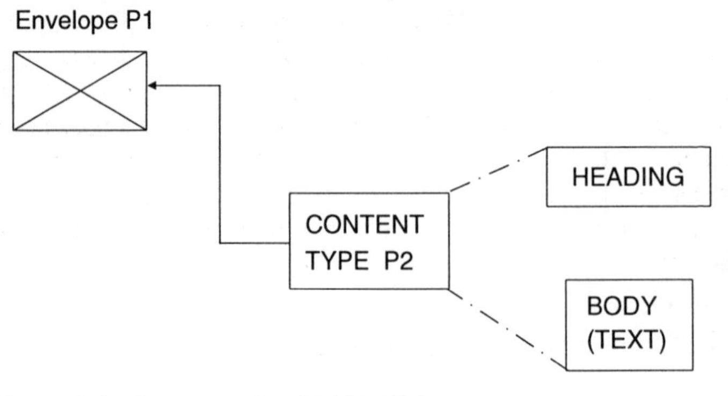

Figure 8.4 Structure of an X.400 IPM

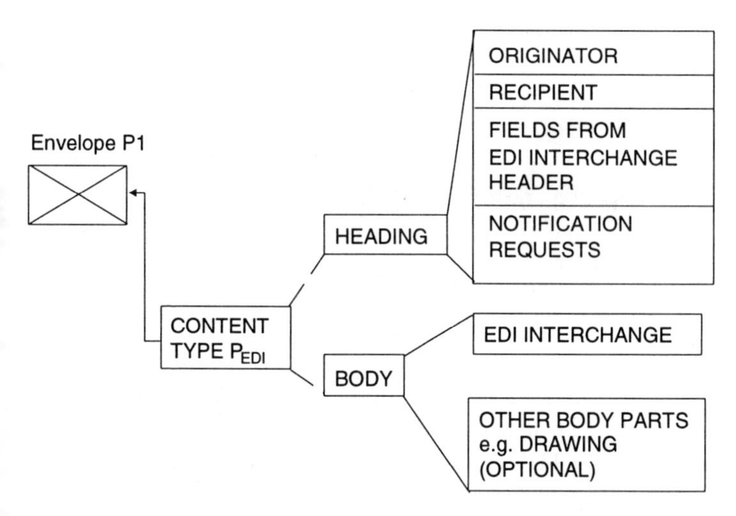

Figure 8.5 P_{edi} and EDIM structures

The second approach, standardised by NIST in the U.S., is to create a new content type, PO, for the EDI message. The EDI message goes directly into the P1 envelope as an 'undefined' content type.

Note that it is not possible to interchange messages of different types between software that supports one type or the other.

The 1988 P_{edi} protocol provides an X.400 message specifically for EDI: the EDIM. A P1 envelope is still used but it has a content type called the 'EDI message type'. X.435 defines a user agent for EDIMs. In contrast to IPM an X.435 user agent does not have a human user but the 'EDI user' is another process, i.e. the computer systems that generate or use EDI data.

The concept of EDI notifications is also supported in 1988 P_{edi}, as in Figure 8.5.

8.5 Internal topologies

The one thing the X.400 and X.435 standards do not specify is how they should be implemented. This leads to a variety of possible solutions, each best suited to a particular business scenario.

Five possible models are listed below but this is by no means an exhaustive list and many combinations of scenarios may be applicable to larger firms. The P_{edi} protocol contains elements in support of all the following:

1. The EDI UA is linked to a single EDI application
2. The EDI UA is a 'corporate gateway', i.e. a sophisticated piece of software capable of audit trails, translations of data format and may onward transmit via proprietary protocols. In this case the gateway may take responsibility for returning notifications of delivery.
3. As above but no responsibility is taken for notification. This will be handled by the final recipient UA.
4. The EDI UA uses a VAN for its MTA services. In this case the VAN would support the MS
5. EDI UA and MTA are co-located entirely on a VAN service. Here the VAN would most likely be supplying not only store and forward services but also protocol translations.

8.6 X.500 directory services

As the use of X.400 increases the need for an automated directory service grows. This was recognised quite soon in the process of X.400 definition and in 1988 a standard for electronic directory services was published and this is X.500 (Wynne, 1990; Riggs, 1994; Communicate, 1995).

X.500 will be particularly useful for EDI since it will allow EDI processes to obtain both ORAddresses of EDI applications on trading partners machines and the specific attributes and capabilities of those applications. This will ensure (if the directory is accurate!) that data can be sent in an appropriate format to an address that is known to be able to handle it.

The structure of X.500 is similar to X.400 in as much as it consists of two key elements: the directory services agent (DSA) and the directory user agent (DUA). A DUA and a DSA converse in a defined protocol called directory access protocol (DAP).

A typical implementation in the EDI scenario is that the DUA will run in the same machine as the EDI UA and requests for information will be passed from the EDI UA to the DUA. The DUA will then extract the information using DAP from a remote directory, perhaps maintained centrally by a VAN (in X.500 terms an ADDMD — administrative directory management domain). The DUA will then pass the information back to the EDI UA which can then format and address the message correctly.

8.7 Security

Security is a major consideration in EDI since much of the data moved using EDI can be termed 'mission critical'. Security is used as a blanket term to cover many different needs according to the type of data and use to which it is being put (Mitchell, 1990).

Three types of security are available with P_{edi}. These can be classified as authentication, data integrity and confidentiality.

A fourth threat to any network is of course availability, however this is generally left to lower level protocols to handle.

Much thought has been given to security in P_{edi}, and indeed X.400 1988 is also much improved in this area. The concept of different levels of security in a network is allowed for and specific elements of the protocol allow a message to be configured for the particular type of security required.

Security in EDI is obtained by the use of keys and encryption. Keys can be of two types, symmetric where the same key is used at both ends and asymmetric where the key used to unlock is different from the key used to lock but has exactly the same effect. One type of asymmetric key algorithm (the RSA) is given in annex to the X.509 standard.

Authentication is about ensuring that the message originates from who it purports to. This is essential where instructions for payment are being issued.

Authentication relies on the application to a message of an asymmetric private key known only to the sender. It works in the following manner.

The message being sent is put through a hashing function and encrypted i.e. locked with the secret key. Both the hashed message and the original message are passed to the receiver who uses a different key to unlock (decrypt) the hashed message.

The unhashed message is compared to the original message. If the results are the same then the message must have come from the originator since he is the only one who could have created the locked message as the key is private.

Data integrity can also be achieved using the above method, but may also be achieved using symmetric keys. Note that symmetric keys can guarantee data integrity but since by definition they are known by at least two people they cannot guarantee authenticity.

Confidentiality can be achieved simply by encrypting the contents of the P1 envelope. The standards do not specify any particular encryption method, which is left free to the implementors of the end systems.

Further security features are typically provided by the VANS. These include audit trails and notifications, both positive and negative.

Full support for notifications is not mandatory, however compliant systems which do not support the accepting of responsibility for a message must forward that responsibility to the next system in the network.

P_{edi} also contains mechanisms to provide proof of content received and non repuditable proof of content received.

Proof of content received is equivalent to returning a copy of the original content to the sender on receipt of a message.

Non repuditable proof allows the originator of a message to be sure the receiver cannot say it was never received. It is similar in concept to the above but the copy sent back can be considered to be notarised as a certificate of origin is passed between the parties.

8.8 Global VAN topologies

With the trend to global marketplaces and global networks three styles of global VAN are evident (Draper, 1994; Dearing, 1995).

In the first case a VAN operates its processing in a single location but uses a network to transport data to it via long lines from all over the world. This approach is the simplest for the VAN but the lack of local processing can make traffic between endpoints in the same remote country uneconomic.

To address this the idea of franchising the service was created. This meant in-country operations and local presence, but international traffic is essentially inter-VAN traffic and often subject to surcharges.

A more up to date method, in line with current trends in distributed processing, is to run a single global service with distributed, co-operating nodes. This is an expensive approach, out of reach for all but the largest players, but it does have significant benefits.

Instead of franchising the service so that each node is an island or long-lining into a single central point the distributed nodes all co-operate. This means that each node may host mailboxes and processing for another to facilitate load balancing or disaster recovery and only a single registration is required for an account to be directly accessible via any node world wide. This allows the global management of a multi-national company's account from the most appropriate regional centre.

The global distributed approach matches both the structure of the standards bodies and the trends in computing which are driving the development of standards such as X.400 and X.500.

8.9 Future trends

The current round of standards work concerns the integration of standards, and this is reflected in current developments. In the EDI world this is the emergence of X.435 as the standard for EDI over X.400. This will allow an organisation to utilise a single physical (OSI) connection to carry all its text messaging whether structured or unstructured. The X.500 directory services will allow the off-net processes to address and route the data to the correct endpoint.

Further work is being done in the area of office document architecture (ODA). This will further integrate the messaging media as it allows a separation of the content of a document from the layout and formatting of the content. Thus it is possible not only to transmit the content of a document but instructions for its presentation. It also allows for multimedia documents so a document which includes a picture can be sent in a single transmission and interpreted correctly at the far end.

In assessing future trends in the VANS we must look at services under development today. Clearly there are three thrusts for the future. These all concern the shift of focus up the OSI stack towards the abstract user orientated functionality. This will protect users' current investment decisions rather than past ones.

The first is the consolidation of vital translation facilities, with the focus moving from the lower levels to the higher.

The second is an extension of the media used to include such things as voice, video etc.

The third is the provision of higher level services. These will be such things as document conversion, directory services and on-net library services where whole document layouts are held on the service but only the content is transmitted.

8.10 References

Communicate (1994a) The electronic postman never knocks, Communicate, July.

Communicate (1994b) Standard procedure needs flexibility and innovation, *Communicate*, July.

Communicate (1995) New applications turn X.500 into a phoenix, *Communicate*, April.

Dearing, B. (1995) EDI: Driving VAN growth, *Telecommunications*, June.

Draper, T. (1994) Public E-Mail: An issue of interoperability, *Telecommunications*, July.

Dunlap, C. (1993) New EDI standard welcomed, *Communications International*, February.

Houldsworth, J. (1990) Applying electronic messaging, *Telecommunications*, August.

Hunt, D. (1995) International messaging on a growth path, *Telecommunications*, April.

Mitchell, C.J. (1990) OSI and X.400 security, *Telecommunications*, May.

O'Brien, A. (1994) Fax and voice messaging technologies: new benefits for E-Mail users, *Telecommunications*, February.

PM (1995) E-Mail checklist for managers, *Professional Manager*, July.

Riggs, B. (1994) X.500: Stuck in the slow lane? *Communications International*, September.

Scales, I. (1993) Is it more trouble than it's worth? *Communications International*, November.

Wynne, G. (1990) X.500 and directories, *Communications*, June.

Young, K. (1993) Let the mail get through, *Network*, March.

9. Telephones and headsets

9.1 Telephones

When we speak our voice sets up sound vibrations which disturb the surrounding air, and travel through the air to be detected by the listener's ear drums. Sound will travel through most media, air, water, wood, plastic, etc. and at different speeds through the different media. For example in air sound travels at approximately 1100 feet per second (335m/s) and approximately 4300 feet per second in water (1311m/s). Electrical signals travel at the speed of light 3×10^8m/s.

The telephone, patented by Alexander Graham Bell in the USA in 1875-77, was an apparatus named the 'Electrical Speaking Telephone'. It was a means of transmitting sound (especially voice) over a distance, by converting sound vibrations into electrical signals which passed through wires as electrical signals, and were then reconverted to sound at the distant end. This will provide one direction of communication. In a practical telephone bothway communication is necessary, so in a simple telephone system each end is provided with means for transmitting and receiving sound.

In addition some form of mechanism is required to signal to the distant end to attract the distant party's attention to the fact that the caller wishes to talk to the distant end. Numerous forms of signalling schemes have been devised and they vary depending on the type of telephone system to which the telephone is connected. Some form of power supply is required to generate the electrical signals and many forms of circuits exist to provide power to the telephone. These consist of local batteries to power only an individual telephone or central batteries in the telephone exchange (Central Office or CO in the USA) where the power is sent down the individual telephone line to the telephone, where circuits exist to extract the power from the line and feed the telephone circuits. Cost considerations have been, and still are, very important in telephone equipment, so the first

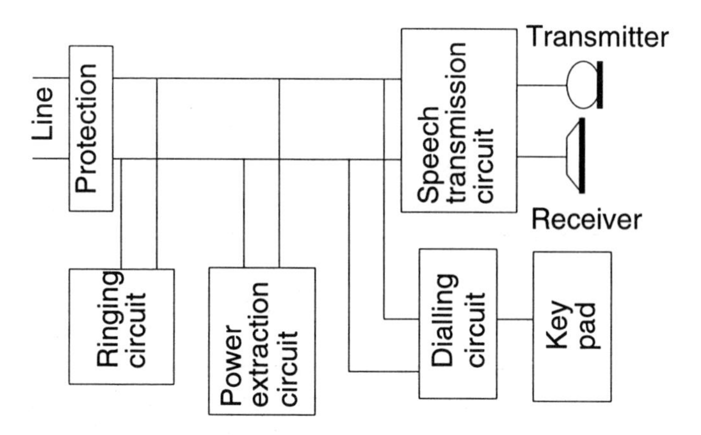

Figure 9.1 Functional diagram of a telephone

practical telephones economised on the wiring from one telephone to the other by combining the pair of wires from the transmitter and from the receiver onto one single pair of wires, to connect both telephones together. The basic functional model for a telephone is as shown in Figure 9.1.

9.2 Telephone speech functions

In speech communications we are primarily concerned with intelligibility i.e. the percentage of voice signals transmitted from one telephone and correctly received at the telephone at the distant end. We do not have to transmit every single part of the sound generated, since the listener acts as an error correction mechanism and can fill in any missing elements and still understand completely what has been transmitted. The prime intelligence in the human voice is contained in quite a small segment of the bandwidth of the hearing/voice spectrum. The human ear can detect sounds from 16Hz to 20000Hz and the human voice can generate sounds from 100Hz to 10000Hz.

Most of the energy in an average male voice is contained within the band from 125Hz to 2000Hz and the average female voice from

400Hz to 2000Hz. ITU-T recommend a bandwidth of 300Hz to 3400Hz as being adequate for telephony and provide an acceptable level of intelligibility on a speech connection.

9.3 Telephone transmitters

Telephone transmitters have been designed using numerous techniques to convert sound impinging onto the transmitter into electrical energy.

9.3.1 Carbon granule transmitter

Carbon granule transmitters were very common from the earliest days of telephony until comparatively recently. The principle of operation is shown in Figure 9.2.

Sound pressure impinges on the diaphragm causing it to vibrate. The centre of the diaphragm is attached to a carbon electrode which moves, compressing and decompressing the fine carbon granules sealed in the chamber at the rear of the device. A second carbon

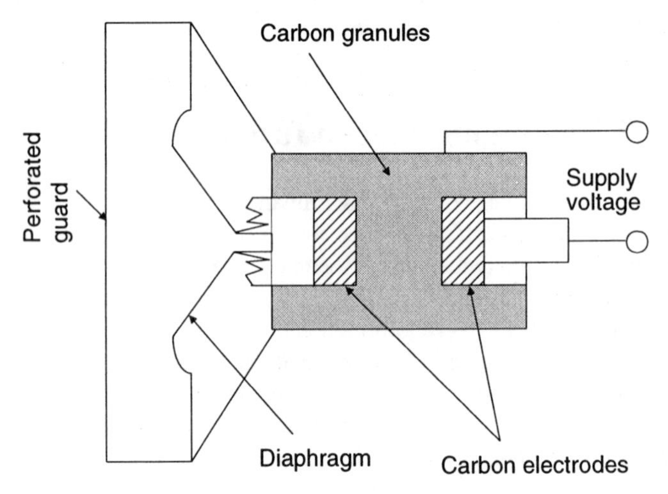

Figure 9.2 Carbon granule transmitter

electrode is fixed to the back wall of the device. The variation in pressure on the fine carbon granules causes the resistance between the two electrodes to vary in unison with movement of the diaphragm. When a d.c. supply is connected between the two carbon electrodes, and the carbon granules are compressed and decompressed, an alternating current is superimposed on the d.c. supply which represents the speech signal.

9.3.2 Rocking armature transmitter

The rocking armature transmitter is shown in Figure 9.3. Sound pressure impinges on the diaphragm causing the armature to pivot, altering the magnetic field and inducing an alternating current in the coils. This alternating current represents the speech signal.

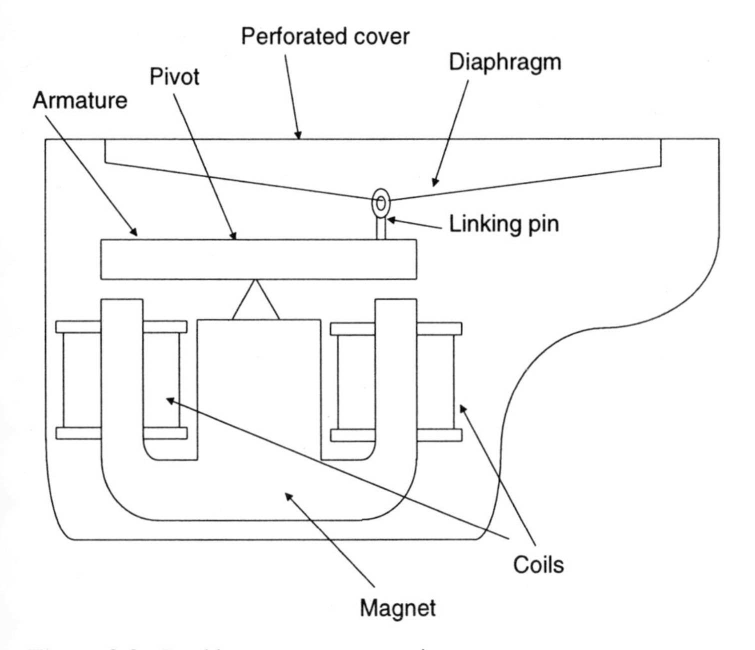

Figure 9.3 Rocking armature transmitter

9.3.3 Piezo-electric transmitter

Piezo-electric transmitters work on the principle that certain types of quartz crystals become electrically polarised when subjected to pressure from sound waves.

Sound impinges on the disc of piezo material, causing the disc to be stressed, varying the charge on the capacitor within the disc, which in turn varies the gate voltage of the FET impedance conversion amplifier, shown in Figure 9.4. The disc and the FET are all housed in a pressed aluminium casing.

The apertures at the front of the transmitter have variable hole patterns which are selected to improve the frequency response of the unit.

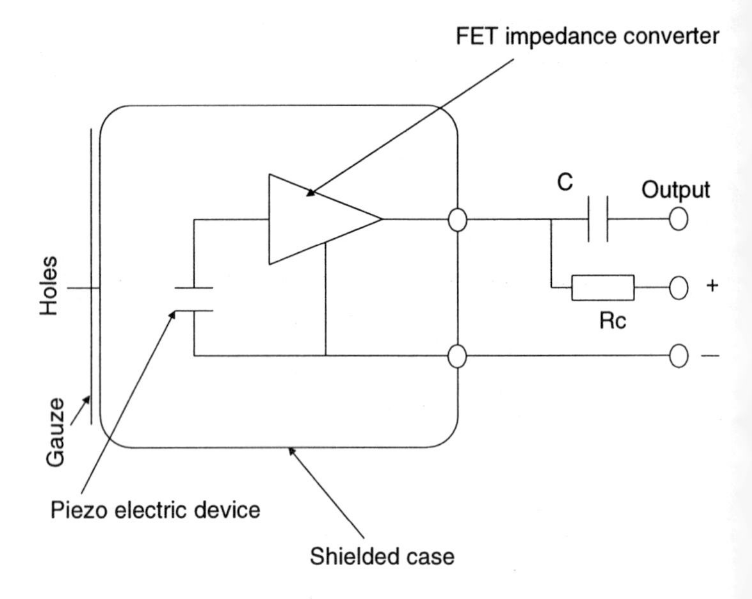

Figure 9.4 Piezo-electric transmitter

9.4 Telephone receivers

Like telephone transmitters, telephone receivers have been designed using numerous principles, but the earliest type tended to be electromagnetic as shown in Figure 9.5.

An electromagnet winding terminates the telephone line and picks up the a.c. signals from the line, varying the magnetic field. The thin iron diaphragm, held in close proximity to the ends of the magnetic poles, is moved to and from the poles of the electromagnet. The greater the current detected in the coils the closer the diaphragm is pulled to the poles of the magnet. The diaphragm therefore vibrates in unison with the received speech signal.

9.4.1 Rocking armature receiver

The construction of this type of receiver is the same as the rocking armature transmitter. When this device is used as a receiver it is

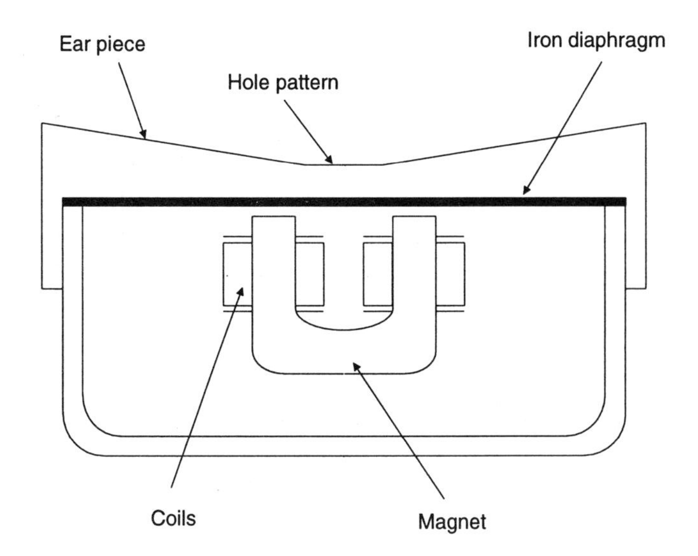

Figure 9.5 Electromagnetic receiver

similar in operation to the electromagnetic type. An a.c. current is detected in the telephone line and this causes the armature to move back and forth in sympathy with the a.c. current on the line. The armature is linked to a diaphragm which vibrates causing sound to be reproduced.

9.4.2 Moving coil receiver

This type of receiver is similar in construction to a loudspeaker, as shown in Figure 9.6. A flexible diaphragm, normally of a plastic material, is formed into a shallow conical shape and a former, wound with very thin wire, is attached to the diaphragm. The assembly is then suspended in the field of a permanent magnet. The coil is connected to the telephone line and detects the a.c. speech current on the line, inducing a change in the magnet field which causes the diaphragm to vibrate, reproducing sound from the received speech signal.

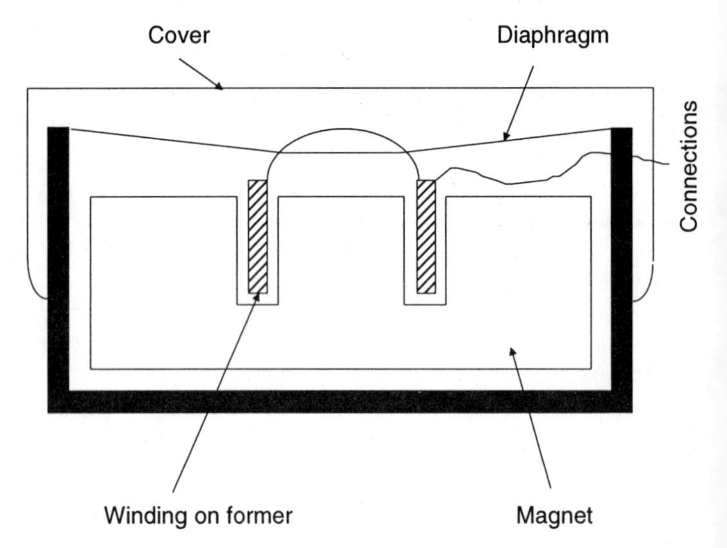

Figure 9.6 Moving coil receiver

9.5 Telephone handset design

The telephone handset is designed to provide the optimum ear/mouth geometry for the average human head. Early designs of telephones used a fixed mouthpiece, mainly due to the early forms of carbon microphone, which were subject to the carbon granules generating unwanted noise when the microphone section of the telephone was moved. It was not until manufacture of the carbon microphone improved that it was possible to design a telephone handset that contained both a transmitter and receiver.

The ergonomics of the handset design are equally important as the acoustic design, to provide the user with a telephone of high performance and reliability of operation. The earpiece needs to be designed to sit comfortably on the user's ear. It must form an acoustic seal with the ear to ensure that received sound is directed into the ear and does not leak away into the surrounding air. If the handset has a poor seal to the ear the listener will pick up external noise and will probably complain that the telephone is not loud enough. Furthermore, others in close proximity to the user will hear the telephone conversation.

The positioning of the microphone is also critical. For reliable performance it must be possible to comfortably position the handset such that with the receiver in place on the ear the microphone sits directly in front of the mouth, to ensure that the microphone is in a direct line of any sound being generated.

The choice of material for the handset and the size and positioning of any holes, is also critical to the acoustic performance of the handset. Mounting of the transmitter and receiver should also be done with care, with sufficient vibration absorbing material placed between the body of the handset and the transmitter or receiver to ensure that sound vibrations do not travel through the handset body from transmitter to receiver or in the other direction and impair performance.

Sidetone is a unique feature of telephone handset design and is essential for good performance of the telephone. When using a telephone handset the normal acoustic feedback from mouth to ear is partially blocked, therefore to compensate a portion of the talker's own voice is fed back into the receiver. Subjective tests have shown that, at high levels of sidetone, the speech power levels transmitted to the telephone network are significantly reduced, resulting in the distant end experiencing a low volume of received sound. This effect is due to the talker's voice being lowered and moving the handset away in an attempt to compensate for the increased sidetone.

High levels of sidetone are also undesirable in environments with high ambient noise since room noise is picked up by the microphone, which is also transmitted to the ear, and this may mask the incoming speech signal, making the telephone unusable.

Acoustic shock is another significant factor in the design of telephone handsets. As the user will have the handset close to the ear during a conversation, it is essential that the telephone circuit has a mechanism to prevent large transients or power surges from the telephone line being transmitted to the receiver, which could produce a sudden surge of sound pressure into the listener's ear-drum.

With modern electronic designs of telephone circuits it is necessary to protect the electronics from static discharge from the user into either the transmitter or receiver, or through the body of the handset where the two sections join. No regulatory specifications exist for electrostatic discharge but 10kV is the minimum and for dry low humidity areas 20kV to 30kV is recommended.

The telephone handset must be balanced so that the user feels comfortable holding the handset for lengthy periods of time. Many manufacturers add weight to the handset. This is not always just for balance since it is sometimes necessary to counteract the 'walk on' problem. If a handset is too light, or the telephone is of a poor mechanical design, the switch to seize the line (hookswitch or cradleswitch) is not returned to its normal position at the end of the call by the weight of the handset, and the exchange equipment remains seized preventing incoming calls arriving.

Finally the telephone handset can take considerable abuse during its lifespan, therefore it is required to withstand severe shock tests, dropping on the floor, being hit against a desk, and still continue to function.

This presents the designer with a considerable problem in meeting all the above diverse requirements to ensure reliable efficient telephone service.

9.6 Transmission performance

To assess the transmission performance of an individual telephone, or telephone network, a number of parameters have been established. The voice operation of a telephone can be broken down into the following elements:

1. The path from talker's mouth to the transmitter.
2. The path from the telephone receiver to the listener's ear.
3. The near end path between the transmitter and the receiver.
3. The telephone connection from transmitter and receiver to the distant end.

Different methods exist for comparing the performance of various telephones and for defining the standard to which a particular telephone or system should perform. The four key parameters in determining telephone performance are:

1. Sending sensitivity.
2. Receiving sensitivity.
3. Impedance the telephone presents to the line.
4. The line impedance for minimum sidetone.

Each of these parameters can be a function of the frequency range of the telephone and the available line current. Algorithms exist for converting the sensitivity of the transmission path or receive path into a loudness rating i.e. a single number for a given frequency, sound pressure level, and line length.

A loudness rating is a standardised method of measuring the transmission loss of a speech path. It is a single value related to the loudness with which the listener perceives, speech that has been emitted by the talker. The performance of a telephone is assessed by measuring between the telephone and an impedance representing the telephone termination, at the telephone exchange, over varying line lengths and over the full frequency range in which the telephone operates. Known sound pressure levels are applied to the microphone of the telephone and the resulting voltage changes appearing at the simulated exchange terminations are measured.

In the other direction signals are applied to the simulated exchange termination and the resulting sound pressure levels are measured at the telephone receiver. These measurements are normally carried out with the telephone in a sound absorbing box or in an anechoic chamber, with the telephone clamped into a jig which has an artificial ear and an artificial voice. These are used to detect and measure the received signal and generate the tones to stimulate the transmitter.

9.6.1 Sending sensitivity

The sending sensitivity is defined as the ratio of the voltage measured at the terminating impedance of the exchange feed bridge and the sound pressure level injected into the microphone. This ratio is normally measured over the full frequency range. The sensitivity depends on the frequency response of the microphone, the acoustic path to the microphone, and the transmit gain of the telephone circuit. The sending sensitivity and frequency response are designed to have a rising characteristic within the speech band. The reason for this is that although most of the power in speech is at the lower frequencies, the sharpness of speech is created by the higher frequencies. However, this is the region where losses in the cable are highest, therefore by increasing the sensitivity at the higher frequencies this tends to compensate for the effect of the cable and increases the clarity of the speech. Above 4kHz the sensitivity should fall off rapidly, to prevent unwanted out of band energy to be transmitted to the network, which might interfere with PCM circuits. At low frequencies the sensitivity should roll off to prevent mains hum pickup. The gain introduced into the sending path is chosen to ensure that the telephone send sensitivity is in accordance with the transmission plan for the telephone network. A typical send sensitivity response is shown in Figure 9.7.

9.6.2 Receive sensitivity

The receive sensitivity is defined as the ratio of the sound pressure level in the artificial ear, to the voltage applied at the terminating

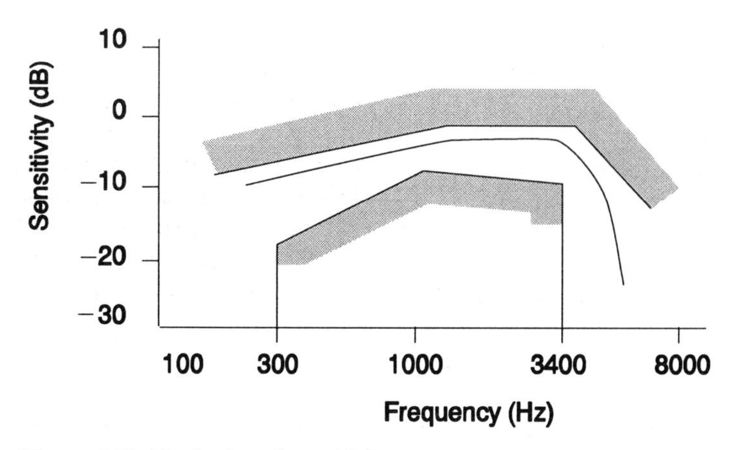

Figure 9.7 Typical send sensitivity response

impedance of the exchange feed bridge. This ratio is also measured over the full frequency range and at different line lengths. The receiving sensitivity is nominally flat within the the 300Hz to 3400Hz speech band and tends to roll off at low frequencies. To reduce the effects of picking up mains hum in the telephone, the cut off at low frequencies cannot be too sharp, since the absence of low frequencies increases the difficulty of hearing on quieter calls.

Above 3400Hz the sensitivity drops off sharply to exclude the unwanted high frequency by-products of pulse code modulation systems. This is because PCM systems sample at 8kHz and all codecs put some part of the sampling frequency onto the line, albeit at a very low level. The aim is to reduce the level as much as possible, to prevent it being an annoyance to the user, even if it is heard. A typical receive sensitivity response is shown in Figure 9.8.

9.6.3 Impedance

The impedance that the telephone should present to the line is chosen to suit the transmission plan for the network. As more digital local exchanges were installed in networks, control of the impedance presented to the network became more important. Digital exchanges

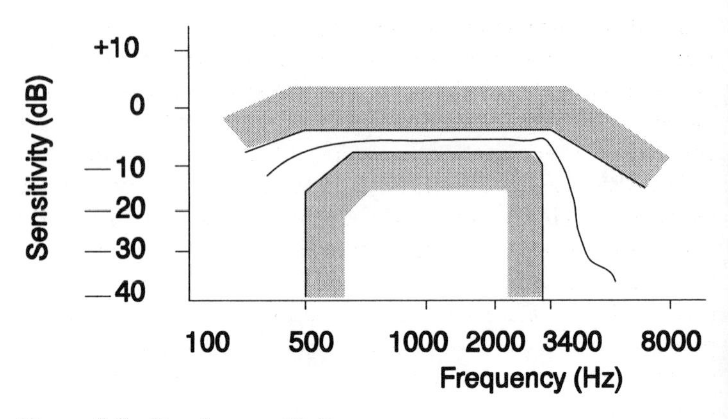

Figure 9.8 Receive sensitivity

use four wire transmission paths – two for transmitting and two for receiving. However, the cable between the exchange and the telephone is only two wire, therefore a mechanism is necessary to terminate the bi-directional two wire path onto the unidirectional send and receive paths of the four wire circuit.

The four wire circuits are balanced to prevent singing or echo and it is also necessary to include the impedance of the line and the telephone to ensure correct balance. Echo is where speakers hear their own voices repeated back. Part of the speech signal energy transmitted in one direction is returned in the other direction due to imperfect balance at the 2/4 wire conversion. The length and type of cable connecting the telephone to the exchange will vary and hence its impedance, therefore trying to match the impedance of the telephone line to the balance network would involve individual measurements of each line. In practice a nominal impedance of 600 ohms has been chosen to terminate the line and most telephones have been designed around this value. More recently complex impedances have been specified by the network provider as this is thought to provide a more ideal balance network.

One essential function of any speech circuit of a telephone is to combine the four wire path from the handset (transmitter connections and receiver connections) onto the two wire path to the exchange line, and to do this with minimum coupling from the transmitter into the receiver (sidetone). Sidetone plays an important part in the subjective performance of any telephone but it is desirable to be able to control the amount of sidetone that can occur over the range of exchange line connections, to ensure that transmission difficulties do not arise.

One way to assess the efficiency of the 2/4 wire converter in the telephone circuit is to check for minimum sidetone, or zero sidetone when the telephone is connected into an impedance which completely suppresses the sidetone. Many telephone speech integrated circuits perform the 2/4 wire conversion by means of a Wheatstone bridge principle, as in Figure 9.9, to obtain proper decoupling between the send and receive signals. For balance of the bridge Equation 9.1 must be satisfied.

$$\frac{R_1}{R_2} = \frac{Z_{line}}{Z_{balance}} \tag{9.1}$$

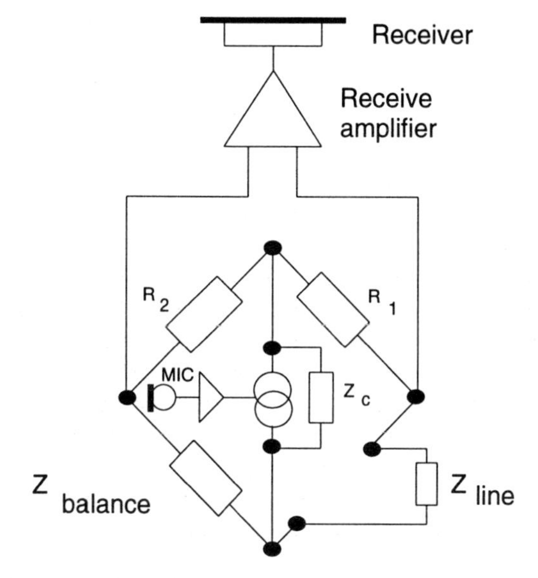

Figure 9.9 Telephone bridge

The a.c. signal from the microphone is fed into one diagonal of the bridge and a small amount is fed into the balance impedance, but the majority is fed to line via R_1. The signal coming from the line is detected on the other diagonal of the bridge then amplified and applied to the receiver.

9.6.4 D.c. characteristics

Most telephones which are directly connected to the exchange via a pair of wires derive their power from a central battery in the exchange, which feeds current down the line normally from a –48V or –50V supply. Some earlier telephone systems operated on local batteries, next to the telephone on the customer's premises, but this method of operation is now only reserved for special applications e.g. radio telephones. The exchange battery feed circuit is usually a Stone or a Hayes bridge. The battery feed has balanced windings with respect to earth and presents a relatively low d.c. resistance, but a high impedance to voice frequencies. The d.c. resistance is usually around 200 ohms in the battery feed and the same in the ground return with an inductance of 2 Henrys.

The d.c. characteristics of a telephone are important as they determine the operating parameters for the telephone. It is normal to have a full bridge rectifier at the front end of any telephone circuit, to guard against line reversals which may damage speech and dialling circuits. When the telephone is idle (on-hook) the telephone will extract current from the line. This is normally only insulation resistance leakage and is of the order of a few micro amperes. However, some telephone network providers permit more current to be taken from the line when the telephone is on-hook, to power electronic memory devices in the telephone or LCD displays.

When the telephone is off-hook current is drawn from the network to power the telephone. The telephone network providers set limits to the amount of current that may be drawn and the telephone must operate within the parameters set. The d.c. characteristics are normally specified by a relationship between current and voltage, as in Figure 9.10.

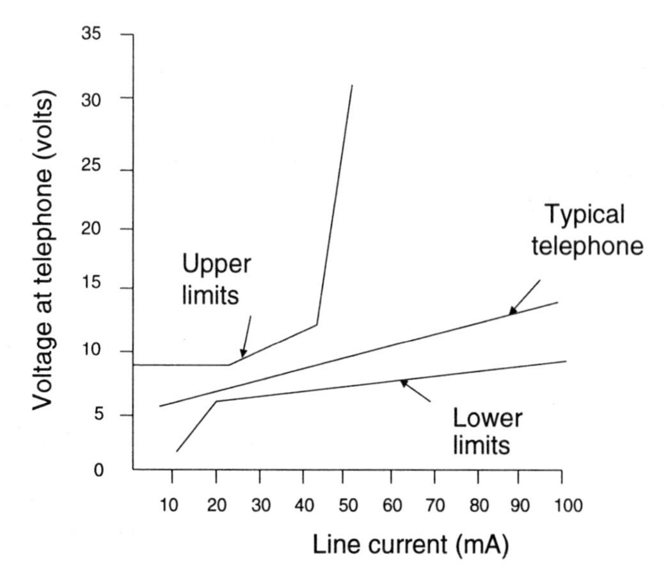

Figure 9.10 D.c. characteristic of a telephone

The upper limits are set to ensure that the telephone will be able to seize and hold the exchange equipment. The off-hook impedance of a telephone should be equal to or greater than the 400 ohms battery feed, to obtain optimum voice signal power transfer in the 200Hz to 3400HZ voice frequency range. Many telecommunication administrations set a lower limit on the voltage that a telephone should operate on, because it is not desirable to subject the feed bridge circuit to a short circuit. However, this is not normally the practical reason since the cable from the exchange must be present. The more practical reason is the need to ensure that sufficient current is available for more than one telephone to operate in parallel.

9.7 Signalling

Signalling in a telephone circuit relates to the ability to seize the exchange in the outgoing state and then pass the address information of the required distant telephone to the exchange. In the incoming state signalling relates to the alerting of the telephone, when it is on-hook, that an incoming call is trying to make contact. A message is then sent to the exchange to inform it that the incoming alerting signal has been successful, as the call has been answered, and the telephone has gone off-hook.

9.7.1 Incoming ringing signals

Ringing is the earliest form of signalling on telephone circuits and has changed little in principle over the years. Even the early manual telephone switchboards used ringing generators that were wound by hand to generate an a.c. signal between 20Hz and 50HZ (70V to 100V). This was superimposed on the exchange battery feed when applied to the line. A.c. signal was then extracted from the line via a capacitor of approximately 2μF to drive a bell. The capacitor also blocked the d.c. battery feed to the telephone, to prevent the d.c. polarising the bell. When the telephone is taken off-hook the bell circuit is short circuited by the telephone receiver and transmitter, resulting in the exchange detecting the change in condition of the line and 'tripping' the ringing circuit.

In today's telephone exchanges the ringing signal is usually a composite of a.c. and d.c. components. The ringing waveform, as shown in Figure 9.11, is typically a sine wave with its axis shifted by the –50V exchange battery. The ringing signal is applied in bursts of a few seconds followed by a period of silence and then repeated.

Some exchange systems are able to send additional signals down the line during the ringing phase to provide extra facilities to the user. The most novel is the use of MODEM tones during the silent phase in US exchange equipment to send calling number identity. This has been made possible by the use of electronic exchange equipment, and the growth in ITU-T No.7 signalling between exchanges, to permit the calling number identity to be transmitted across the network.

In the telephone the traditional device to alert the caller has been a bell, consisting of an electromagnet with two coils and an armature which strikes one or two gongs when operated. Modern telephones use a variety of electro-acoustic devices to generate the ringing sound.

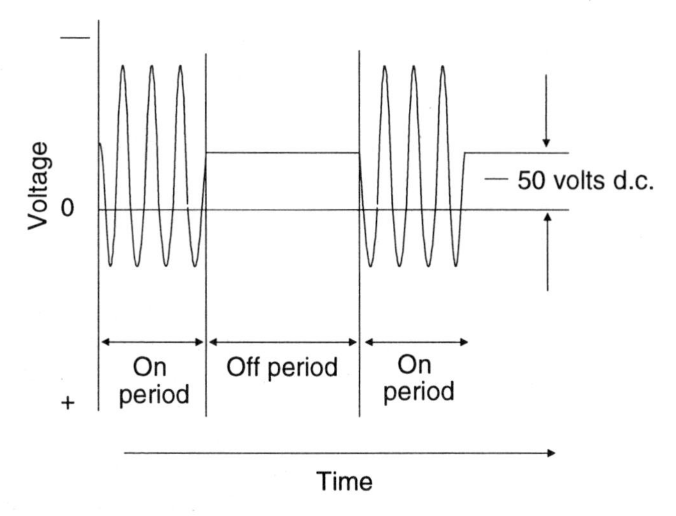

Figure 9.11 Typical ringing waveform

One of the cheapest forms of device is the ceramic disc resonator. This consists of a brass disc onto which is deposited a ceramic coating. When a potential is applied between the two surfaces the disc resonates at its fundamental frequency. To drive these devices in a telephone circuit the coupling to the line uses the conventional capacitor followed by a full bridge rectifier, to rectify the a.c. ringing signal, then to an IC which amplifies or restricts the signal to a level suitable for the ceramic disc. The ceramic disc is housed inside the plastic moulding of the telephone in an acoustic chamber to ensure good sound reproduction.

9.7.2 Outgoing signalling

Each telephone in the network is assigned a unique number and each country has been assigned a unique number, therefore to obtain a distant telephone it is necessary to pass this number (address) to the telephone exchange, which will then route the request through the network to the distant telephone. There are two basic types of address signalling:

1. Dial pulse or loop disconnect.
2. Multifrequency.

9.7.3 Dial pulse or loop disconnect signalling

Dial pulse or loop disconnect signalling takes its name from the rotary dial method of signalling information to the exchange. The rotary dial is a mechanical device which is connected across the telephone line with a pair of contacts. The user places a finger in the dial and turns the dial round to the finger stop and then releases the dial. The dial then returns to its original position by a spring, which was wound up when the dial was turned. On its return the dial speed is held constant by a mechanical governor and a cam with notches makes and breaks the telephone line, sending a train of pulses down the line representing the number dialled.

When the dial is first moved by the user off the normal rest position, a further set of contacts are activated (off normal contacts) to short circuit the speech circuit in the telephone, and prevent the user hearing clicks.

Dial pulses are used to signal the address to the exchange by a series of pulses at 10 pulses per second with a tolerance of ±1 pulse as the design objective. The make break ratio is also important for the type of exchange the telephone is signalling into. The ratio is either 60% or 67% nominal break period. The interdigital pause is the period at the end of pulsing as the dial returns to its rest position and on most dials this is approximately 600ms to 800ms. A typical dial pulse train is shown in Figure 9.12.

Modern telephones use integrated circuits to generate the dial pulses. The IC will be connected to a keyboard, and so will have a keyboard scanning circuit and key debouncing circuits, to ensure correct registration of the keys selected.

The IC will also provide facilities for muting the speech circuits during dialling, for much the same reason as the off normal contacts in the mechanical dial. In addition to the basic dialling functions the addition of memory cells to the IC provides the user with functions such as last number redial and stores for most frequently dialled numbers.

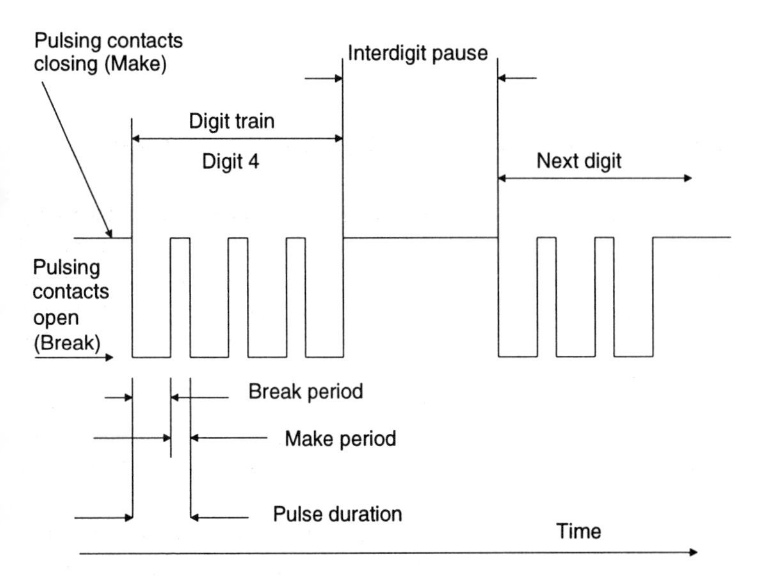

Figure 9.12 Dial pulse train

9.7.4 Dual tone multifrequency signalling

Multifrequency signalling consists of a combination of two tones with frequencies within the speech band, each combination of two frequencies representing a single digit. The tones are presented on the line for approximately 100ms and then an intertone pause of 100ms before the next combination of tones is sent. The speed of dialling is therefore many times faster than loop disconnect dialling. The tones are in two groups, a low band and a high band, and they are geometrically spaced to ensure that any two frequencies of a valid combination are not harmonically related. Table 9.1 details the frequencies and valid combinations.

The ICs are designed to ensure that frequencies do not deviate more than ±1.5% from the standard and that no unwanted signals are sent to line. In addition the speech transmission is normally muted during signalling, but the receiver normally receives some of the tone to give the user confidence that the tone is being sent to line. The ICs

Table 9.1 DTMF tone assignments

Low	*High*			
	1209Hz	*1336Hz*	*1447Hz*	*1633Hz*
697Hz	1	2	3	A
770Hz	4	5	6	B
852Hz	7	8	9	C
941Hz	*	0	#	D

normally have the same facilities for last number redial and digit storage as explained earlier.

DTMF signalling has the added advantage of end to end signalling. Once the connection is set up DTMF signals can be transmitted from the telephone in-band over the link to the distant to control voice mail machines, answering machines and access banking services, etc.

9.8 Loudspeaking telephones

Loudspeaking telephones offer a variety of different functions to the user and in this category of telephone we have the following facilities:

1. Call progress monitor.
2. Group listening.
3. Full handsfree operation.

A telephone with call progress monitoring facilities is equipped with a small loudspeaker and a monitor amplifier. The user presses a button to activate the telephone, seizing the line, and dial tone is applied to the line from the exchange equipment and then fed through the monitor amplifier. The user can dial the distant end, listen to ringing tone and when the distant end answers must then pick up the handset to be able to speak as the only transmitter available is in the handset.

The group listening telephone normally has call progress monitor, but in addition has a circuit to enable the user to press a button on the telephone and broadcast the incoming speech signal to others listening in the room. The circuit may also drive the receiver in the handset simultaneously with the loudspeaker in the base of the telephone. Outgoing speech is only possible through the handset transmitter.

The full handsfree telephone is equipped with both a handset and a separate microphone and loudspeaker for handsfree operation. The microphone and speaker are now all provided as part of the telephone, although some older designs consisted of separate units connected in parallel with the standard telephone.

The design of the handsfree telephone creates particular problems. The level of sound being emitted from the loudspeaker is required to be large enough to fill the volume of air in the room where it is being used, so that users can be anywhere in the room and still hear the incoming speech. The microphone in the telephone will also need to pick up this incoming speech signal from the acoustic feedback. Therefore, to prevent the telephone from howling, attenuators need to be switched in to the microphone path. These circuits are often referred to as antilarsen circuits.

Figure 9.13 shows a block diagram of typical handsfree system. When a handsfree telephone connection is established speech signals can be reflected at the two to four wire interface, either in the local exchange or in the digital PBX two to four wire interface, as well as round the local sidetone loop from loudspeaker to microphone. Hence oscillations and speech distorting can occur and in the worst case howling.

To have full duplex working on the loudspeaking telephone requires very fast switching between the transmit and receive circuits and compensation for background noise. This requirement has led to the development of microprocessor controlled handsfree circuits, where the control algorithms are stored in the microprocessor.

Particular attention must be paid to the acoustic design of the telephone body for handsfree operation. The loudspeaker should be housed in a speaker box and acoustically isolated from the micro-

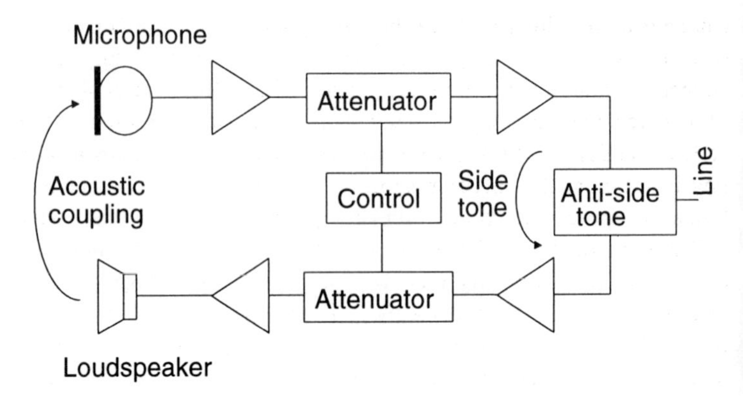

Figure 9.13 Handsfree telephone

phone as far as possible, to prevent sound travelling through the body of the telephone to the microphone and causing instability. Sound emanating from the back of the speaker can be particularly problematical and many telephones have some sound absorbing material behind the speaker. Microphones for loudspeaking telephones tend to be much more sensitive than telephone transmitters in handsets as they are required to detect speech coming from a few feet away in the room.

9.9 Digital telephones

Digital telephones are characterised by having the codec in the telephone to convert the speech from analogue to digital and vice versa. The PCM encoded speech is then sent to line at a rate of 64kbit/s. In addition a 16kbit/s signalling channel is also present to send address and alerting signals to the exchange. ITU-T have now defined standard interfaces for digital operation in the I series of recommendations, but many proprietary interfaces exist.

The I series specifies a 192kbit/s full duplex digital path between the digital telephone (terminal) located in the user's premises and the local exchange. The 192kbit/s is divided into a B1 channel (64kbit/s), a B2 channel (64kbit/s) and a D channel (16kbit/s). The B channels can be used for either voice or data. A digital telephone will consist of a handset and possibly handsfree loudspeaker and microphone connected to an integrated circuit, which will provide the functions of send, receive gain, frequency response, sidetone generation and the codec functions. Additional circuits are necessary to provide the physical drive to the line, extract power from the line and operate the link access procedure on the D channel (LAPD).

The functionality of the telephone can be built into the microprocessor software (functional model) or respond to stimulus signalling from the exchange (stimulus model). In the former all the call processing logic is designed into the telephone, whilst in the latter it is held centrally in the exchange and the telephone responds only to the messages from the exchange.

9.10 Telephone standards

European standards for telephones are country specific and contain many national peculiarities, which prevent the production of a single harmonised telephone suitable for use anywhere in Europe. NET 4 (Norme Europeen de Telecommunication) for basic PSTN access was produced under an EEC directive in 1986 and covers all the essential parameters of signalling, transmission and safety.

A new EEC directive in 1991 established the concept of CTRs (Common Technical Regulations) which will supersede the NETs with the aim of establishing the essential requirements to be regulated. The harmonised standards give the specifications and tests to be applied to ensure conformity with the requirements.

The CTRs will not duplicate other requirements already in existence, such as the LVD (Low Voltage Directive). The objective is, some time in the future, to be able to test and approve telephones in one country in Europe, apply the EC Mark, and sell that telephone in any other country without further approval testing. At present each country issues its own national standards for telephones, either via the PTT as the controlling authority or via a national standards organisation such as BSI (British Standards Institute) in the UK.

In North America the position on standards for telephones is much simpler. The FCC (Federal Communications Commission) issues the basic standards for all of the USA and in Canada the DOC (Department of Communications) issues similar standards.

9.11 Headsets

Rapid growth in the use of computers and telemarketing means that more and more people in many walks of life are finding themselves before a visual display unit, operating a keyboard accessing data, and using a telephone terminal with a headset. Headsets, once the domain of the telephone switchboard operator, are playing the major role in solving their problem. With the continuing growth in this area it is vital that telephone system designers, manufacturers and suppliers co-operate closely with specialist telecommunication headsets companies early on in the design stage to ensure that the customer gets sufficient choice of quality products.

With the ACD market sector growing at an ever increasing rate, organisations large and small are having to face the fact that more and more of their employees need to have their hands free to operate a computer keyboard when conducting telephone conversations.

Companies involved in the growing business of telemarketing and financial services, and those using direct order entry, such as express delivery services, airline reservations, as well as such public service organisations as police, fire and ambulance are all making use of headsets to overcome the problem, and at the same time increasing their employee's productivity and efficiency.

9.12 Headset aesthetics

Most headsets today are light and efficient and offer excellent sound quality. It is important, however, that operators using these sets, often for most of the day, should look right and feel right when wearing them. When a headset is worn, there is a tendency to regard it as an article of clothing, and if the user does not like the headset then he or she will find a reason for not wearing it. So aesthetics as well as comfort are of prime importance.

Headsets are frequently perceived to be expensive but, in comparison with the cost of the telephone systems they are used with, they represent but a fraction of the cost. OEMs need to recognise at the design stage that the headset is the all important man-machine interface, and although an accessory to the system, its performance is a critical aspect of the total system design.

The recently introduced range of quality headsets, such as Profile, are of advanced design and construction and offer exceptional performance. They use the latest surface mount technology and because of their micro-miniature design, and the need for customisation to a wide range of PABX, ACD, Key Systems and other telephone instruments, are generally hand-assembled. The flex cord, too, needs to be of high quality to ensure reliability and, in some cases, is tested to withstand 750000 test flexures.

Compared with a handset, the headset has to be light yet able to withstand hard use. It is constantly picked up and frequently dropped and tends to be under constant flexure.

Developments in modern plastics, and the introduction of the electret microphone have allowed development of headsets weighing ounces and of spider-like proportions, yet which are amazingly strong and flexible. There are even plastics which allow the headset to be formed to fit the head simply by heating with friction from the hand. Headsets come in a variety of forms, over-the-head, under-the-chin, on-the-ear, but the market is tending to standardise on two major types: the over-the-head and the on-the-ear versions. The tip-in-ear variety is in decline, partly due to the fact that it can irritate the inside of the ear, as well as cause aural hygiene problems. Hygiene is also an important consideration in respect of ease of cleaning and maintaining.

9.13 Headset technical considerations

The ergonomics and mechanical design of modern headsets are of paramount importance. Different styles and configurations of headsets, perhaps with a modular concept, are essential. Limited choice of headsets places suppliers and communication managers in an unnecessarily inadequate position.

The acoustic and electrical performance of the headset clearly is also fundamental to its design. In particular, the microphone and earphone transducers must be matched correctly, both to interface with the system being used and to suit user requirements. Ideally, the microphone and earphone transducers in the headset need to operate to the same electrical and acoustic performance as the handset. This is not always easy to achieve mainly due to size and weight considerations, and often sound pressure levels need to be compensated for by automatic gain control and adjustable volume control.

9.13.1 Microphones

Most headsets use either standard electret or electromagnetic microphones. Both types can be boom mounted in miniature form, but the electret type is generally most popular. Acoustic tubes are used on some headsets but these generally provide poorer transmission quality than boom mounted microphones and are subject to considerable crosstalk problems.

Noise cancelling versions of both electret and electromagnetic microphones are also available. However the operation of noise cancelling microphones can often degrade transmission performance if users are not trained properly. There is a tendency to presume that a noise cancelling microphone will automatically provide a superior transmission performance. This is not the case. To explain this it is necessary to know how a noise cancelling microphone works.

The definition of a noise cancelling microphone is a microphone which functions according to the pressure gradient principle. It has two sound ports which give direct access to the front and the back of the microphone diaphragm. The sound ports are symmetrical around the diaphragm.

The noise cancelling microphone reacts to the difference of sound pressure on the diaphragm. It is necessary to distinguish between the sound field originating from a close sound source (called 'near field') and the sound field originating from a distant sound source ('far field').

Noise is a typical distant sound source. The microphone reacts differently to the two types of sound source. The near field frequency response closely resembles a 'normal' pressure microphone characteristic, while the far field frequency response shows a distinct proportional dependence on the frequency.

The directional characteristic of noise cancelling microphones is such that greater attenuation of microphone signals takes place when the sound activates the microphone simultaneously on each side of the diaphragm. In a polar form the characteristic would be shaped as a figure of eight. This means there are two noise reducing effects present in a noise cancelling microphone, partly the effect of direction and partly a strong dampening of low frequencies from the far field sound source.

The near field measurement less the far field measurement is termed noise cancellation. A good noise cancelling microphone has a noise cancellation of not less than 11dB.

Experience with use of noise cancelling microphones is that there is a very minimal real requirement for this type of microphone except in specialised uses such as military, heavy industrial, air traffic control etc. The main objection from other users being the distance dependence of the microphone i.e. the need for careful positioning of the microphone in close proximity to the lips of the user.

9.13.2 Earphones

Handsets normally have a high acoustic impedance and high gain output. Most headsets cannot achieve the same acoustic impedance and gain without amplification. With close coupled earphones, as used on over the head style and under the chin style headsets, this can be achieved easily either with or without an amplifier, according to the system specification with which they are interfacing. However, to overcome the use of unhygienic eartips on single sided headsets, an acoustic low impedance earphone is used which has been specifically designed to work with an acoustic leak between the earphone diaphragm and the ear canal.

The first headset to employ this principle is GN Netcom's Profile which has an adjustable earphone to provide users with maximum comfort combined with optimum sound pressure output from the earphone.

9.14 The growing need for headsets

PABX and ACD system designers have always taken care to account for the needs of the headset user. Now more and more key system and telephone instrument designers are also starting to recognise those needs particularly in view of the increasing telemarketing activities in the UK.

In the US, every telephone instrument and system normally has a facility to operate with a headset. And since the UK tends to follow closely trend in the USA, we can expect that this will happen here. With screen based work booming, telephone system designers must ensure that all new equipment, including 2-wire sets, have a headset port as defined in BS 6301.

To facilitate optional handset/headset working a switch arrangement can be incorporated simply into many systems. The Danaswitch provided by GN Netcom for this purpose features in many key systems currently on the market.

There is also general feeling among leading headset designers that if OEMs worked with them at an early stage in the design of equipment, and there are headset companies perfectly willing and able to do this, then users would benefit considerably from increased versatility and a greater choice of headsets.

9.15 Headset approval process

In terms of approvals, much depends upon the telephone system that the headset is to be used with, and it is frequently the OEM who obtains the required approval, under DTI specification 85/013.

The approval process is a lot easier nowadays since the introduction of the BABT simple attachment scheme. Previously, testing was carried out under the direct control of BABT and, in the event of failure, the modified apparatus had to repeat the entire test procedure, at considerable expense. Under the new arrangement, testing is carried out directly with one if the accredited test houses who then submit their report to BABT. Testing, and charging, is carried out until a failure occurs or the test is completed. In the event of a failure, the test house informs the designer who can then take the equipment away, rectify it and bring it back to continue testing at the point at which failure occurred.

The test is designed to ensure that:

1. The headset does not electrically damage the public telephone system.
2. That it meets electrical safety standards.
3. That no damage to the user can occur through acoustic shock.

9.16 Headset design criteria

First and foremost it must be comfortable. If it is required to be worn for long periods, then weight is vital too. There are few headsets which can be tolerated continuously over an eight hour day. So it is essential to know the shift period that the user will work.

Its design must be attractive to the user.

Does the user have to move around the telephone point in doing the job? If so, a quick-release connector is essential: it also reduces wear and tear on the terminal plug.

Monaural receivers work well when the user needs to listen in two places at once, e.g. a front desk receptionist/telephonist. Binaural receivers help shut out background noise.

Is there a need for split listening? e.g. radio in one ear, telephone in the other, with the ability to switch the microphone from one to the other.

Boom-mounted microphones are better than acoustic tubes and essential in noisy environments.

Under-the-chin headsets do not disturb the hair. Over-the-head style headsets are stable and more robust. Contour 'on-the-ear' style headsets are light, less noticeable. Conversion kits are an important new cost development and allow users to experiment with both over-the-head and on-the-ear styles to decide their preferences. Over-the-ear pads are comfortable, will not cause infection and block out some external sound. In-the-ear tips offer high close acoustic coupling. Some people may find the tips to be uncomfortable, irritating and unhygienic.

The environment is all important. In hot climates, ears sweat. Vinyl ear pads can be wiped clean with a sterilising pad, while foam pads absorb perspiration and need replacing regularly.

Is the office noisy? Does it lack acoustic damping, are the surfaces hard, is there a lot of paper shuffling? In particularly noisy environments, noise cancelling microphones are available but they must be positioned properly for optimum performance. If not, their use can be counter productive.

Automatic Gain Control (AGC) on the microphone circuit cuts the gain when the user is not talking. A manual gain control on the headset allows adjustment to suit line conditions and the speech level of individual callers. A thumbwheel control is usually preferred to a switchable control. Preferably this should provide for –3dB to +12dB gain adjustment.

Modular design eases logistics for maintenance purposes.

After sales service is vitally important. Headsets are fragile and an efficient and speedy after-sales back-up is essential to most users.

9.17 References

Alexander, J. (1990) A right earful, *Communicate*, October.

Andersen, D. et al. (1989) Sandard subsets, *Electrical Communications, 63, (1).*

Costello, J. (1989) Engaging telephones, *Communications*, July.

Gattey, P. (1993) The 'shocking' truth about headsets, *TE&M*, 1 February.

Hatano, H. and Hara, H. (1990) A digital public telephone for ISDN, *NTT Review*, March.

Rudvin, S. (1990) Towards the ISDN terminal, *Telecommunications*, November.

Timms, S. (1988) ISDN CPE — the way ahead, *Telecommunications*, July.

Vesely, L.I. and Chicoine, F.J. (1991) Handset testing gets human touch, *TE&M*, 1 December.

10. Acronyms

Every discipline has its own 'language' and this is especially true of telecommunications, where acronyms abound. In this guide to acronyms, where the letters within an acronym can have slightly different interpretations, these are given within the same entry. If the acronym stands for completely different terms then these are listed separately.

ACD	Automatic Call Distribution. (Facility for allowing incoming calls to be queued and distributed to waiting service operators.)
ACK	Acknowledgement. (Control code sent from a receiver to a transmitter to acknowledge the receipt of a transmission.)
ADDF	Automatic Digital Distribution Frame. (System used to replace manual distribution frames in plesiochronous transmission networks.)
ADM	Add-Drop Multiplexer. (Term sometimes used to describe a drop and insert multiplexer.)
ADSL	Asymmetrical Digital Subscriber Loop. (Technique for providing broadband over copper.)
AE	Anomaly Events. (E.g. frame errors, parity errors, etc. ITU-T M.550 for digital circuit testing.)
AFC	Automatic Frequency Control.
AFRTS	American Forces Radio and Television Services.
AGC	Automatic Gain Control.
AIN	Advanced Intelligent Network. (Bellcore released specification for provision of wide range of telecommunication capabilities and services.)
AIS	Automatic Intercept System. (System which is programmed to provide information automatically to a

	telephone caller, who has been intercepted and routed to it.)
ALS	Alternate Line Service. (For example two different directory numbers accociated with one line.)
ALU	Arithmetic Logic Unit.
AMA	Automatic Message Accounting. (Ability within an office to record call information automatically for accounting purposes. See also CAMA.)
ANI	Automatic Number Identification. (Feature for automatically determining the identity of the caller.)
APNSS	Analogue Private Network Signalling System.
AP	Application Processes. (Processes within computer based systems which perform specified tasks.)
API	Application Programming Interface.
ARS	Automatic Route Selection. (Facility in an equipment, usually a PABX or multiplexer, to select automatically the best route for transmission through a network.)
ASCII	American Standard Code for Information Interchange. (Popular character code used for data communications and processing. Consists of seven bits, or eight bits with a parity bit added.)
ASIC	Application Specific Integrated Circuit. (Integrated circuit components which can be readily customised for a given application.)
ASTIC	Anti-SideTone Induction Coil. (Hybrid transformer used in voice transmission systems.)
BABT	British Approvals Board for Telecommunications.
BASIC	Beginners' All-purpose Symbolic Instruction Code. (Computer programming language.)
BCC	Block Check Character. (A control character which is added to a block of transmitted data, used in checking for errors.)
BCD	Binary Coded Decimal. (An older character code set, in which numbers are represented by a four bit sequence.)

BDF	Building Distribution Frame.
BELLCORE	Bell Communications Research. (Research organisation, incorporating parts of the former Bell Laboratories, established after the divestiture of AT&T. Funded by the BOCs and RBOCs to formulate telecommunication standards.)
BIST	Built In Self Test.
BHCA	Busy Hour Call Attempts. (It is the number of calls placed in the busy hour.)
BISYNC	Binary Synchronous communications. (Older protocol used for character oriented transmission on half-duplex links.)
BORSCHT	Battery, Overload protection, Ringing, Supervision, Coding, Hybrid, and Test access. (These are the functions provided in connection with a subscriber line circuit. The functions are usually implemented by an integrated circuit.)
BSGL	Branch System General Licence. (Telecommunications licence in the UK.)
BSI	British Standards Institute.
CAS	Channel Associated Signalling. (ITU-T signalling method.)
CASE	Computer Aided Software Engineering (or Computer Aided System Engineering).
CBX	Computer Controlled PBX.
CCB	Coin Collection Box. (Pay phone.)
CCC	Clear Channel Capacity.
CCD	Charged Coupled Device. (Semiconductor device used for analogue storage and imaging.)
CCIR	Comite Consultatif Internationale des Radiocommunications. (International Radio Consultative Committee. Former standards making body within the ITU and now part of its new Radiocummunication Sector.)

CCIS	Common Channel Interoffice Signalling. (North American signalling system which uses a separate signalling network between switches.)
CCITT	Comite Consultatif Internationale de Telephonique et Telegraphique. (Consulative Committee for International Telephone and Telegraphy. Standards making body within the ITU, now forming part of the new Standardisation Sector.)
CCR	Commitment, Concurrency and Recovery.
CCS	Common Channel Signalling. (ITU-T standard singalling system. Also called CCSS.)
CCSS	Common Channel Signalling System. (ITU-T standard signalling system. Also called CCS or Number 7 signalling.)
CDDI	Copper Distributed Data Interface. (Name given to FDDI running over copper media.)
CDR	Call Detail Recording. (Feature within a PABX for call analysis.)
CENELEC	Comite Europeen de Normalisation Electrotechnique. (Committee for European Electrotechnical Standardisation.)
CEPT	Conference des administrations Europeenes des Postes et Telecommunications. (Conference of European Posts and Telecommunications administrations. Body representing European PTTs.)
CICC	Contactless Integrated Circuit Card. (ISO standard for a smart card in which information is stored and retrieved without use of conductive contacts.)
CIT	Computer Integrated Telephony.
CLAN	Cableless Local Area Network. (Radio based LAN.)
CLI	Command Line Interface. (Usually refers to an interface which allows remote asynchronous terminal access into a network management system. Also referred to as Command Line Interpreter.)
CLI	Calling Line Identity. (Facility for determining identity of the caller. See also CLID.)

CLID	Calling Line Identification. (Telephone facility which allows the called party to determine the identity of the caller.)
CMS	Call Management System.
CMT	Character Mode Terminal. (E.g. VT100, which does not provide graphical capability.)
CO	Central Office. (Usually refers a central switching or control centre belonging to a PTT.)
CoCom	Co-ordinating Committee on Multilateral Export Controls.
COCOT	Consumer Owned Coin Operated Telephone. (Privately owned coin boxes linking to public telephone lines.)
CODEC	COder-DECoder.
CPE	Customer Premise Equipment.
CP/M	Control Programme for Microcomputers. (Operating system popularly used for microcomputers.)
CPU	Central Processing Unit. (Usually part of a computer.)
CR	Carriage Return. (Code used on a teleprinter to start a new line.)
CRA	Call Routing Apparatus.
CRC	Cyclic Redundancy Check. (Bit oriented protocol used for checking for errors in transmitted data.)
CRT	Cathode Ray Tube.
CS	Central Station.
CSBS	Customer Support and Billing System.
CSTA	Computer Supported Telephony Applications. (Or Computer Supported Telecommunication Applications. For example, telemarketing applications.)
CSU	Channel Service Unit. (Subscriber line terminating unit in North America.)
CT	Cordless Telephony. (CT1 is first generation; CT2 is second generation, and CT3 is third generation.)
CTR	Common Technical Regulation. (European Community mandatory standard.)

CTS	Conformance Test Services. (Part of European Commission initiative.)
CTS	Clear To Send. (Control code used for data transmission in modems.)
DASS	Digital Access Signalling System. (Signalling system introduced in the UK prior to ITU-T standards I.440 and I.450.)
DASS	Demand Assignment Signalling and Switching unit.
DATTS	Data Acquisition Telecommand and Tracking Station.
DBMS	Database Management System.
DCE	Data Circuit termination Equipment. (Exchange end of a network, connecting to a DTE. Usually used in packet switched networks.)
DCF	Data Communications Function.
DCN	Data Communications Network.
DCR	Dynamically Controlled Routing. (Traffic routing method proposed by Bell Northern Research, Canada.)
DDC	Data Country Code. (Part of an international telephone number.)
DDD	Direct Distance Dialling. (Generally refers to conventional dial-up long distance calls placed over a telephone network without operator assistance.)
DDF	Digital Distribution Frame. (Frame for physical connection of transmission lines. See also ADDF and MDDF.)
DDI	Direct Dialling In. (External caller able to dial directly to an extension.)
DDI	Distributed Data Interface. (Proposal to run the FDDI standard over unshielded twisted pair.)
DDN	Digital Data Network.
DDN	Defence Data Network. (US military network, derived from the ARPANET.)
DDP	Distributed Data Processing.

DDS	Digital Data Service. (North American data service.)
DE	Defect Events. (E.g. loss of signal, loss of frame synchronisation, etc. ITU-T M.550 for digital circuit testing.)
DECT	Digital European Cordless Telephony. (Or Digital European Cordless Telecommunications. ETSI standard, intended to be a replacement for CT2.)
DID	Direct Inward Dialling. (PABX feature allowing an external caller to connect to an extension without first going through an operator.)
DIN	Deutsches Institue fur Normung. (Standardisation body in Germany.)
DMA	Direct Memory Access.
DN	Directory Number. (Of a customer in a switching system.)
DNHR	Dynamic Non-Hierarchical Routing. (Traffic routing method implemented by AT&T.)
DNIC	Data Network Identification Code. (Part of an international telephone number.)
DOD	Department Of Defence. (US agency.)
DOV	Data Over Voice. (Technique for simultaneous transmission of voice and data over telephone lines. This is a less sophisticated technique than ISDN.)
DPNSS	Digital Private Network Signalling System. (Inter-PABX signalling system used in the UK.)
DS-0	Digital Signal level 0. (Part of the US transmission hierarchy, transmitting at 64kbit/s. DS-1 transmits at 1.544Mbit/s, DS-2 at 6.312Mbit/s, etc.)
DSAP	Destination Service Access Point. (Refers to the address of service at destination.)
DSB	Double Sideband.
DSBEC	Double Sideband Emitted Carrier.
DSC	District Switching Centre. (Part of the switching hierarchy in BT's network.)
DSE	Data Switching Exchange. (Part of packet switched network.)

DSI	Digital Speech Interpolation. (Method used in digital speech transmission where the channel is activated only when speech is present.)
DSP	Digital Signal Processing.
DSS	Digital Subscriber Signalling. (CCIT term for the N-ISDN access protocol.)
DSU	Data Service Unit. (Customer premise interface to a digital line provided by a PTT.)
DSX-1	Digital Signal Crossconnect. (Crossconnect used for DS-1 signals.)
DTE	Data Terminal Equipment. (User end of network which connects to a DCE. Usually used in packet switched networks.)
DTMF	Dual Tone Multi-Frequency. (Telephone signalling system used with push button telephones.)
DTP	Distributed Transaction Processing.
DXC	Digital Crossconnect. (See also DCX and DCS.)
EBCDIC	Extended Binary Coded Decimal Interchange Code. (Eight bit character code set.)
EC	European Commission.
ECMA	European Computer Manufacturers Association.
ECSA	Exchange Carriers Standards Association. (USA)
ECU	European Currency Unit. (Monetary unit of the EEC, created in 1981.)
EDI	Electronic Data Interchange. (Protocol for interchanging data between computer based systems.)
EDIFACT	EDI For Administration Commerce and Transport. (International rules for trading documents, e.g. purchase orders, payment orders, etc.)
EDP	Electronic Data Processing.
EEA	Electrical Engineering Association.
EEA	Electronic and business Equipment Association.
EEC	European Economic Community.
EEMA	European Electronic Mail Association.

EEPROM	Electrically Erasable Read Only Memory. (Integrated circuit used to store data, which can be erased by electrical methods.)
EET	Equipment Engaged Tone. (Tone customer receives when there is no free line for the call.)
EFT	Electronic Funds Transfer.
EFTA	European Free Trade Association.
EFTPOS	Electronic Funds Transfer at the Point Of Sale.
EIA	Electronic Industries Association. (Trade association in USA)
EMA	Electronic Mail Association.
E-MAIL	Electronic Mail.
EMC	Electromagnetic Compatibility.
EMI	Electromagnetic Interference.
EN	Equipment Number. (Code given to a line circuit, primarily in switches, to indicate its location on equipment racks.)
EN	Europaische Norm. (Norm Europeenne or European standard.)
ENV	European pre-standard.
EOT	End Of Transmission. (Control code used in transmission to signal the receiver that all the information has been sent.)
EOTC	European Organisation for Testing and Certification.
EOTT	End Office Toll Trunking. (US term for trunks which are located between end offices situated in different toll areas.)
EOW	Engineering Order Wire. (A channel for voice or data communication between two stations on a transmission line.)
EP	European Parliment.
ESN	Electronic Serial Number. (Usually refers to the personal identity number coded into mobile radio handsets.)
ESO	European Standardisation Organisation.

ESPRIT	European Strategic Programme for Research and Development in Information Technologies.
ESS	Electronic Switching System. (A generic term used to describe stored programme control exchange switching systems.)
ETE	Exchange Terminating Equipment.
ETIS	European Telecommunications Information Services. (Part of CEPT.)
ETNO	European Telecommunications Network Operators. (Association of European public operators.)
ETS	Electronic Tandem Switch.
ETS	European Telecommunication Standard. (Norme Europeenne de Telecommunications. Standard produced by ETSI.)
ETSI	European Telecommunications Standards Institute.
EUCATEL	European Conference of Associations of Telecommunication industries.)
FAS	Frame Alignment Signal. (Used in the alignment of digital transmission frames.)
FAX	Facsimile.
FCC	Federal Communications Commission. (US authority, appointed by the President to regulate all interstate and international telecommunications.)
FCS	Frame Check Sequence. (Field added to a transmitted frame to check for errors.)
FDM	Frequency Division Multiplexing.
FDDI	Fibre Distributed Digital Interface. (Standard for optical fibre transmission.)
GATT	General Agreement on Tariffs and Trade.
GDN	Government Data Network. (UK private data network for use by government departments.)
GEN	Global European Network. (Joint venture between European PTOs to provide high speed leased line and switched services. Likely to be replaced by METRAN in mid 1990s.)

GoS	Grade of Service. (Measure of service performance as perceived by the user.)
GOSIP	Government OSI Profile. (Government procurement standard.)
GSC	Group Switching Centre. (Part of the hierarchy of switching in BT's network. Also called the primary trunk switching centre by ITU-T.)
GSLB	Groupe Special Large Bande. (A CEPT broadband working group.)
GVPN	Global Virtual Private Network.
HCI	Human Computer Interface.
HD	Harmonisation Document. (Sometimes used to describe an EN.)
HOMUX	Higher Order Multiplexer.
HRDS	Hypothetical Reference Digital Section. (ITU-T G.921 for digital circuit measurements.)
HSE	Health and Safety Executive. (UK)
HSSI	High Speed Serial Interfacr. (Specification for high data rate transmission over copper cable.)
IA2	International Alphabet 2. (Code used in a teleprinter, also called the Murray code.)
IA5	International Alphabet 5. (International standard alphanumeric code, which has facility for national options. The US version is ASCII.)
IAB	Internet Activities Board.
IBC	Integrated Broadband Communications. (Part of the RACE programme.)
IBCN	Integrated Broadband Communications Network.
ICA	International Communications Association. (Telecommunications users' group in the USA.)
ICAO	International Civil Aviation Organisation.
ICMP	Internet Control Message Protocol. (Protocol developed by DARPA as part of Internet for the host to communicate with gateways.)

IDA	Integrated Digital Access. (ISDN pilot service in the UK.)
IDDD	International Direct Digital Dialling.
IDN	Integrated Digital Network. (Usually refers to the digital public network which uses digital transmission and switching.)
IEC	International Electrotechnical Commission.
IEC	Interexchange Carrier. (US term for any telephone operator licensed to carry traffic between LATAs interstate or intrastate.)
IEEE	Institute for Electrical and Electronics Engineers. (USA professional organisation.)
IETF	Internet Engineering Task Force.
I-ETS	Interim European Telecommunicaitons Standard. (ETSI.)
IF	Intermediate Frequency.
IFIPS	International Federation of Information Processing Societies.
IN	Intelligent Network.
INTUG	International Telecommunications Users' Group.
IO	International Organisation.
I/O	Input/output. (Usually refers to the input and output ports of an equipment, such as a computer.)
IP	Internet Protocol.
IPC	Inter-Processor Communications.
IPC	International Private Circuit.
IPM	Inter-Personal Messaging. (Use of Electronic Data Interchange.)
IPR	Intellectual Property Rights.
IPVC	International Private Virtual Circuit.
IS	International Standard.
IS	Intermediate System.
ISD	International Subscriber Dialling.
ISDN	Integrated Services Digital Network. (Technique for the simultaneous transmission of a range of services, such as voice, data and video, over telephone lines.)

ISI	Inter-Symbol Interference. (Interference between adjacent pulses of a transmitted code.)
ISLAN	Integrated Services Digital Network. (LAN which can carry an integrated service, such as voice, data and image.)
ISM	Industrial, Scientific and Medical. (Usually refers to ISM equipment or applications.)
ISO	International Standardisation Organization.
IT	Information Technology. (Generally refers to industries using computers e.g. data processing.)
ITA	International Telegraph Alphabet.
ITAEGT	Information Technology Advisory and co-ordination Expert Group on private Telecommunication network standards.
ITC	International Trade Commission.
ITSTC	Information Technology Steering Committee. (Comite de Direction de la Technologie de l'Information. Consists of CEN/CENLEC/ETSI.)
ITU	International Telecommunication Union.
ITU-T	International Telecommunication Union Telecommunication sector.
ITU-R	International Telecommunication Union Radiocommunication sector.
ITUSA	Information Technology User Standards Association.
IVA	Integrated Voice Application.
IVDS	Interactive Video and Data Services.
IVDT	Integrated Voice and Data Terminal. (Equipment with integrated computing and voice capabilities. In its simplest form it consists of a PC with telephone incorporated. Facilities such as storage and recall of telephone numbers is included.)
IVM	Integrated Voice Mail.
IVR	Interactive Voice Response.
IXC	Interexchange Carrier. (USA long distance telecommunication carrier.)

JEIDA	Japan Electronic Industry Development Association.
JESI	Joint European Standards Institute. (CEN and CENELEC combined group in the area of information technology.)
JIS	Japanese Industrial Standard. (Product marking used in Japan to denote conformance to a specified standard.)
JISC	Japanese Industrial Standards Committee. (Standards making body which is funded by the Japanese government.)
JSA	Japanese Standards Association.
JTM	Job Transfer and Manipulation. (Communication protocols used to perform tasks in a network of interconnected open systems.)
KBS	Knowledge Based System.
kHz	KiloHertz. (Measure of frequency. Equals to 1000 cycles per second.)
LAMA	Localised Automatic Message Accounting. (The ability to use AMA within a local office.)
LAN	Local Area Network. (A network shared by communicating devices, usually on a relatively small geographical area. Many techniques are used to allow each device to obtain use of the network.)
LAP	Link Access Protocol.
LAPB	Link Access Protocol Balanced. (X.25 protocol.)
LAPD	Link Access Protocol Digital. (ISDN standard.)
LAPM	Link Access Protocol for Modems. (ITU-T V.42 standard.)
LASER	Light Amplification by Stimulated Emission of Radiation. (Laser is also used to refer to a component.)

LATA	Local Access and Transport Area. (Area of responsibility of local carrier in USA. When telephone circuits have their start and finish points within a LATA they are the sole responsibility of the local telephone company concerned. When they cross a LATA's boundary, i.e. go inter-LATA, they are the responsibility of an interexchange carrier or IEC.)
LBA	Least Busy Alternative. (Traffic routing strategy defined for fully connected networks.)
LBO	Line Build Out. (The extension of the electrical length of a line, for example by adding capacitors.
LBRV	Low Bit Rate Voice. (Speech encoding technique which allows voice transmission at under 64kbit/s.)
LCD	Liquid Crystal Display.
LCN	Local Communications Network. (ITU-T)
LCR	Least Cost Routing. (Usually applies to the use of alternative long distance routes, e.g. by using different carriers, in order to minimise transmission costs.)
LDDC	Long Distance D.C. line signalling. (Method of d.c. signalling.)
LEC	Local Exchange Carrier. (USA local telecommunication carrier.)
LIFO	Last In First Out. (Technique for buffering data.)
LISN	Line Impedance Stabilising Network. (An artificial network used in measurement systems to define the impedance of the mains supply.)
LJU	Line Jack Unit.
LLC	Logical Link Control. (IEEE 802. standard for LANs.)
LLP	Lightweight Presentation Protocol.
LSB	Least Significant Bit. (Referring to bits in a data word.)
LSI	Large Scale Integration. (Describes a complex integrated circuit semiconductor component.)
LSTR	Listener SideTone Rating. (Measure of room noise sidetone effect.)

LTE	Line Terminating Equipment. (Also called Line Terminal Equipment. Equipment which terminates a transmission line.)
MA	Multiple Access.
MAC	Media Access Control. (IEEE standard 802. for access to LANs.)
MAN	Metropolitan Area Network.
MAP	Manufacturing Automation Protocol.
MASER	Microwave Amplification by Simulated Emission of Radiation.
MAT	Metropolitan Area Trunk. (A cable system which is used to reduce crosstalk effects in regions where there is a large number of circuits between exchanges.)
MCM	Multichip module.
MCS	Maritime Communication System.
MDDF	Manual Digital Distribution Frame.
MDF	Main Distribution Frame.
MDNS	Managed Data Network Service. (Earlier proposal by CEPT which has now been discontinued.)
MEP	Member of the European Parliament.
METRAN	Managed European Transmission Network. (CEPT initiative to provide a broadband backbone across Europe.)
MF	Multi Frequency. (A signalling system used with push-button telephones.)
MFJ	Modification of Final Judgement. (Delivered by Judge Harold Greene. 1982 Act in the AT&T divestiture case.)
MHS	Message Handling System. (International standard.)
MHz	MegaHertz. (Measure of frequency. Equal to one million cycles per second.)
MICR	Magnetic Ink Character Recognition.
MIPS	Million Instructions Per Second. (Measure of a computer's processing speed.)

MITI	Ministry of International Trade and Industry. (Japanese.)
MITT	Minutes of Telecommunication Traffic. (Measure used by telecommunication operators for tariffing purposes.)
MMI	Man Machine Interface. (Another name for the human-computer interface or HCI.)
MoD	Ministry of Defence. (UK)
MODEM	Modulator/Demodulator. Device for enabling digital data to be send over analogue lines.
MOTIS	Message Oriented Text Interchange System. (ISO equivalent of a message handling system.)
MoU	Memorandum of Understanding.
MPT	Ministry of Posts and Telecommunications. (Japan.)
MS	Management System.
MSC	Main Switching Centre. (Part of the switching hierarchy in BT's network.)
MSC	Mobile Switching Centre. (Switching centre used in mobile radio systems.)
MSDOS	Microsoft Disk Operating System. (Very popular operating system for PCs.)
MSI	Medium Scale Integration. (Usually refers to an integrated circuit with a medium amount of on-chip circuit density.)
MTBF	Mean Time Between Failure. (Measure of equipment reliability. Time for which an equipment is likely to operate before failure.)
MTN	Managed Transmission Network.
MTS	Message Transfer System.
MTS	Message Telephone Service. (US term for a long distance telephone service.)
MTTR	Mean Time To Repair. (A measure of equipment availability. It is the time between an equipment failure and when it is operational again.)
NANP	North American Numbering Plan. (Telephone numbering scheme administered by Bellcore.)

NATA	North American Telecommunications Administration.
NBS	National Bureau of Standards. (USA.)
NDC	National Destination Code. (Part of numbering system.)
NE	Network Element.
NET	Nome Europeenne de Telecommunication. (European Telecommunications Standard, which is mandatory.)
NFAS	Not Frame Alignment Signal. (In transmitted code.)
N-ISDN	Narrowband Integrated Services Digital Network.
NIST	National Institute for Standards and Technology. (USA.)
NLC	Network Level Control. (Or Network Layer Control.)
NMC	Network Management Centre.
NMI	Network Management Interface. (Term used within OSI to indicate the interface between the network management system and the network it manages.)
NOI	Notice Of Inquiry. (FCC paper for comment.)
NPA	Numbering Plan Area.
NPI	Null Point Indication.
NPP	Network Performance Parameters.
NSAP	Network Service Access Point. (Prime address point used within OSI.)
NT	Network Termination. (Termination designed within ISDN e.g. NT1 and NT2.)
NTE	Network Terminating Equipment. (Usually refers to the customer termination for an ISDN line.)
NTIA	National Telecommunications Industry Administration. (USA)
NTN	Network Terminal Number. (Part of an international telephone number.)
NTT	Nippon Telegraph and Telephone. (Japanese carrier.)
NTU	Network Terminating Unit. (Used to terminate subscriber leased line.)

OA&M	Operations, Administration and Maintenance. (Also written as OAM.)
OAM&P	Operations, Administration, Maintenance and Provisioning.
O&M	Operations and Maintenance.
OATS	Open Area Test Site. (Used for EMC measurements.)
OEM	Original Equipment Manufacturer. Supplier who makes equipment for sale by a third party. The equipment is usually disguised by the third party with its own labels.)
OFTEL	Office of Telecommunications. (UK regulatory body.)
OLR	Overall Loudness Rating. (Measurement of end to end connection for transmission planning.)
ONI	Operator Number Identification. (Operator used in a CAMA office to obtain verbally the calling number for calls originating in offices not equipped with ANI.)
ONP	Open Network Provision.
OSI	Open Systems Interconnection. (Refers to the seven layer reference model.)
OSIE	OSI Environment.
OSP	Operator Service Provider. (Company in the USA providing competitive toll operator services for billing and call completion.)
PABX	Private Automatic Branch Exchange. (PBX in which automatic connection is made between extensions.)
PAD	Packet Assembler/Disassembler. (Protocol converter used to provide access into the packet switched network.)
PANS	Peculiar And Novel Services. (Often used in conjunction with POTS.)
PBX	Private Branch Exchange. (This is often used synonymously with PABX.)

PC	Personal Computer.
PC	Private Circuit.
PCM	Pulse Code Modulation. (Transmission technique for digital signals.)
PCMCIA	Personal Computer Memory Card International Association.
PIN	Personal Identification Number. (Security number used for items such as remote database entry.)
PLC	Plant Level Controller.
PLL	Phase Locked Loop. (Technique for recovering the clock in transmitted data. Often performed by an integrated circuit.)
PMBX	Private Manual Branch Exchange. (PBX with connections between extensions done by an operator.)
POP	Point Of Presence. (Local access point into network, such as for the Internet. Also refers to the point of change over of responsibility from the local telephone company, within a LATA, to the long distance or inter-LATA carrier.)
POTS	Plain Old Telephone Service. (A term loosely applied to an ordinary voice telephone service.)
PPL	Phase Locked Loop. (Component used in frequency stability systems such as demodulators for frequency modulation.)
PRA	Primary Rate Access. (ISDN, 30B+D or 23B+D code.)
PRBS	Pseudo Random Binary Sequence. (Signal used for telecommunication system testing.)
PRF	Pulse Repetition Frequency. (Of a pulse train.)
PROM	Programmable Read Only Memory. (Memory technology, usually semiconductor, where the data is written, or programmed, once by the user and thereafter can only be read and not changed. See also ROM.)
PSDN	Packet Switched Data Network. (Or Public Switched Data Network. X.25 network, which may be private or public.)

PSPDN	Packet Switched Public Data Network.
PSE	Packet Switching Exchange.
PSN	Packet Switched Network.
PSN	Public Switched Network.
PSS	Packet Switched Service. (Data service offered by BT.)
PSTN	Public Switched Telephone Network. (Term used to describe the public dial up voice telephone network, operated by a PTT.)
PTN	Public Telecommunications Network.
PTO	Public Telecommunication Operator. (A licensed telecommunication operator. Usually used to refer to a PTT.)
PTT	Postal, Telegraph and Telephone. (Usually refers to the telephone authority within a country, often a publicly owned body. The term is also loosely used to describe any large telecomunications carrier.)
PUC	Public Utility Commission. (In USA.)
PVC	Permanent Virtual Circuit. (Method for establishing a virtual circuit link between two nominated points. See also SVC)
QA	Q interface Adaptor.
QAF	Q Adaptor Function.
QD	Quantising Distortion.
QoS	Quality of Service. (Measure of service performance as perceived by the user.)
RACE	Research and development in Advanced Communication technologies in Europe.
RAM	Random Access Memory
RBER	Residual Bit Error Ratio. (Measure of transmission quality. ITU-T Rec. 594-1.)
RBOC	Regional Bell Operating Company. (US local carriers formed after the divestiture of AT&T.)
RCU	Remote Concentrator Unit.
RDA	Remote Database Access.

RDBMS	Relational Database Management System.
RDN	Relative Distinguishing Name. (Used within OSI network management.)
RDSQL	Relational Database Structured Query Language.
RFI	Radio Frequency Interference.
RFNM	Ready For Next Message.
ROM	Read Only Memory. (Memory device, usually semiconductor, in which the contents are defined during manufacture. The stored information can be read but not changed. See also PROM.)
ROS	Remote Operation Service.
ROTL	Remote Office Test Line. (Technique for remotely testing trunk circuits.)
ROW	Right Of Way. (Usually refers to costs associated with laying cables.)
RTNR	Real Time Network Routing. (Dynamic traffic routing strategy being implemented by AT&T.)
RTS	Request To Send. (Handshaking routine used in anlogue transmission, such as by modems.)
RTSE	Reliable Transfer Service Element.
RTT	Regie des Telegraphes et des Telephones. (Belgian PTT.)

S×S	Step by Step. (Refers to a Strowger electromechanical switch.)
SAC	Special Access Code. (Special telephone numbers e.g. 800 service.)
SAP	Service Access Point. (Port between layers in the OSI seven layer model, one for each of the layers, e.g. LSAP, NSAP, etc.)
SCADA	Supervisory Control and Data Acquisition. (Interface for network monitoring.)
SCC	Standards Council of Canada.
SCVF	Single Channel Voice Frequency. (One of the signalling systems used in telex, i.e. ITU-T R.20.)

SDR — Speaker Dependent Recognition. (Speech recognition technique which requires training to individual caller's voice.)
SDU — Service Data Unit. (Data passed between layers in the OSI Seven Layer model.)
SDXC — Synchronous Digital Crossconnect.
SELV — Safety Extra Low Voltage circuit. (A circuit which is protected from hazardous voltages.)
SHF — Super High Frequency.
SIM — Subscriber Identity Module. (Usually a plug in card used with a mobile radio handset.)
SIO — Scientific and Industrial Organisation.
SIP — Societa Italiana per l'Esercizio delle Telecomunicazion. (Italian PTT.)
SIR — Speaker Independent Recognition. (Speech recognition technique which does not need to be trained to individual caller's voice.)
SITA — Societe Internationale de Telecommunications Aeronautiques. (Refers to the organisation and its telecommunication network which is used by many of the World's airlines and their agents, mainly for flight bookings.)
SLC — Subscriber Line Charge. (USA term for flat charge paid by the end user for line connection.)
SLIC — Subscriber Line Interface Card. (Circuitry which provides the interface to the network, usually from a central office switch, for digital voice transmission.)
SM — Service Module.
SMDR — Station Message Detail Recording. (Feature in a PABX for call analysis.)
SMDS — Switched Multimegabit Data Service. (High speed packet based standard proposed by Bellcore.)
SNI — Subscriber Network Interface.
SNMP — Simple Network Management Protocol. (Network management system within TCP/IP.)
SNR — Signal to Noise Ratio.

SNV	Association Suisse de Normalisation. (Swiss standards making body.)
SOHO	Small Office Home Office (market).
SONET	Synchronous Optical Network. (Synchronous optical transmission system developed in North America, and which has been developed by ITU-T into SDH.)
SP	Service Provider.
SPC	Stored Program Controller. (Usually refers to a digital exchange.)
SPEC	Speech Predictive Encoding Communications.
SQL	Structured Query Language.
SSB	Single Sideband.
SSBSC	Single Sideband Suppressed Carrier modulation. (A method for amplitude modulation of a signal.)
SSMA	Spread Spectrum Multiple Access.
STD	Subscriber Trunk Dialling.
STE	Signalling Terminal Equipment.
STMR	SideTone Masking Rating. (Measure of talker effects of sidetone.)
STP	Shielded Twisted Pair. (Cable.)
STP	Signal Transfer Point.
STS	Space-Time-Space. (Digital switching method.)
STX	Start of Text. (Control character used to indicate the start of data transmission. It is completed by a End of Text character, or ETX.)
SVC	Switched Virtual Circuit. (Method for establishing any to any virtual circuit link. See also PVC.)
TA	Telecommunication Authority.
TA 84	Telecommunications Act of 1984. (UK.)
TASI	Time Assignment Speech Interpolation. (Method used in analogue speech transmission where the channel is activated only when speech is present. This allows several users to share a common channel.)

TC	Transport Class. (E.g. TC 0, TC 4, etc.)
TCM	Time Compression Multiplexing. (Technique which separates the two directions of transmission in time.)
TDM	Time Division Multiplexing.
TE	Terminal Equipment.
TEDIS	Trade Electronic Data Interchange System. (European Commission programme.)
TEI	Terminal Endpoint Identifier. (Used within ISDN Layer 2 frame.)
TELMEX	Telefonos de Mexico. (Mexican PTT.)
TELR	Talker Echo Loudness Rating. (Overall loudness rating of the talker echo path.)
TEM	Transverse Electromagnetic cell. (Code used for measuring characteristics of receivers such as pagers.)
TEMA	Telecommunication Equipment Manufacturers Association. (UK.)
TIA	Telecommunication Industry Association. (US based. Formed from merger of telecommunication sector of the EIA and the USTSA.)
TINA	Telecommunication Information Network Architecture.
TM	Trade Mark.
TMA	Telecommunication Managers' Association. (UK)
TMN	Telecommunications Management Network.
TNV	Telecommunication Network Voltage circuit. (Test circuit for definition of safety in telecommunication systems.)
TR	Technical Report. (ISO technical document; not a standard.)
TSI	Time Slot Interchange. (Switching system technique which switches between circuits by separating the signals in time.)
TST	Time-Space-Time. (Digital switching method.)
TTC	Telecommunications Technology Committee. (Japanese.)

TTE — Telecommunication Terminal Equipment.

TTY — Teletypewriter. (Usually refers to the transmission from a teletypewriter, which is asynchronous ASCII coded.)

TUA — Telecommunications Users' Association. (UK.)

TUF — Telecommunications Users' Foundation. (UK.)

TWX — Teletypewriter exchange service. (Used in Canada.)

TeraFLOP — (Trillion Floating Point Operations per second. (Measure of a super computer.)

UAP — User Application Process.

UART — Universal Asynchronous Receiver/Transmitter. (The device, usually an integrated cirucit, for transmission of asynchronous data. See also USRT and USART.)

UDF — Unshielded twisted pair Development Forum. (Association of suppliers promoting transmission over UTP.)

UDP — User Datagram Protocol.

UHF — Ultra High Frequency. (Radio frequency, extending from about 300MHz to 3GHz.)

UI — User Interface.

UL — Underwriters Laboratories. (Independent USA organisation involved in standards and certification.)

UN — United Nations.

UNCTAD — United Nations Conference on Trade And Development.

UNI — User Network Interface. (Also called User Node Interface. External interface of a network.)

UPS — Uninterrupted Power Supply. (Used where loss of power, even for a short time, cannot be tolerated.)

UPT — Universal Personal Telecommunications. (ITU-T concept of the personal telephone number.)

USART — Universal Synchronous/Asynchronous Receiver/Transmitter. (A device, usually an integrated circuit, used in data communication devices, for

	conversion of data from parallel to serial form for transmission.)
USB	Upper Sideband.
USITA	US Independent Telephone Association.
USRT	Universal Synchronous Receiver/Transmitter. (A device, usually an integrated circuit, which converts data for transmission over a synchronous channel.)
USTA	US Telephone Association.
USTSA	US Telecommunication Suppliers Association.
UTP	Unshielded Twisted Pair. (Cable.)

VAD	Voice Activity Detection. (Technique used in transmission systems to improve bandwidth utilisation.)
VADS	Value Added Data Service.
VAN	Value Added Network
VANS	Value Added Network Services.
VAS	Value Added Service. (See also VANS.)
VASP	Value Added Service Provider.
VBR	Variable Bit Rate.
VCO	Voltage Controlled Oscillator. (Component used in frequency generating systems.)
VDU	Visual Display Unit. (Usually a computer screen.)
VF	Voice Frequency. (Signalling method using frequencies within speech band. Also called in-band signalling. Also refers to the voice frequency band from 300Hz to 3400Hz.)
VFCT	Voice Frequency Carrier Telegraph.
VHF	Very High Frequency. (Radio frequency in the range of about 30MHz and 300MHz.)
VHSIC	Very High Speed Integrated Circuit.
VLF	Very Low Frequency. (Radio frequency in the range of about 3kHz to 30kHz.)
VLSI	Very Large Scale Integration. (A complex integrated circuit.)
VMS	Voice Messaging System.

VNL	Via Net Loss. (The method use to assign minimum loss in telephone lines in order to control echo and singing.)
VOA	Voice Of America.
VPN	Virtual Private Network. (Part of a network operated by a public telephone operator, which is used as a private network.)
WACK	Wait Acknowledgement. (Control signal returned by receiver to indicate to the sender that it is temporarily unable to accept any more data.)
WAN	Wide Area Network.
WARC	World Administrative Radio Conference. (Of ITU.)
WATTC	World Administrative Telephone and Telegraph Conference. (Of ITU.)
WBLLN	Wideband Leased Line Network.
WIMP	Windows, Icons, Mouse and Pointer. (Display and manipulation technique for graphical interfaces, e.g. as used for network management.)
WMO	World Meteorological Organisation.
WS	Work Station.
WSF	Workstation Function.
WTO	World Trade Organisation.
WWW	World Wide Web. (Internet.)
ZVEI	Zuverein der Elektronisches Industrie.
ZZF	Zentrasamt fur Zulassungen im Fernmeldewesen.

Index